AF576057

Managing Water Resources in Large River Basins

Managing Water Resources in Large River Basins

Editors

William Young
Nagaraja Rao Harshadeep

MDPI • Basel • Beijing • Wuhan • Barcelona • Belgrade • Manchester • Tokyo • Cluj • Tianjin

Editors

William Young
World Bank
USA

Nagaraja Rao Harshadeep
World Bank
USA

Editorial Office
MDPI
St. Alban-Anlage 66
4052 Basel, Switzerland

This is a reprint of articles from the Special Issue published online in the open access journal *Water* (ISSN 2073-4441) (available at: https://www.mdpi.com/journal/water/special_issues/water_resources_river_basins).

For citation purposes, cite each article independently as indicated on the article page online and as indicated below:

LastName, A.A.; LastName, B.B.; LastName, C.C. Article Title. *Journal Name* **Year**, *Volume Number*, Page Range.

ISBN 978-3-0365-0466-7 (Hbk)
ISBN 978-3-0365-0467-4 (PDF)

Cover image courtesy of William Young.

Contents

About the Editors

William Young (Dr.) is a Lead Water Resources Management Specialist with the World Bank. He is the Program Manager for the Central Asia Water and Energy Program and leads water resource management and irrigation lending operations in Central and South Asia. He has global oversight of the World Bank's multiple national water security studies. From 2013–2016 he was the Program Manager for the South Asia Water Initiative that supports transboundary water cooperation in the Himalayan river basins of South Asia. Roles prior to joining the World Bank include Director of the CSIRO Water Flagship in Australia, and Director of Basin Modeling at the Murray-Darling Basin Authority. He is also currently an Adjunct Professor in Water Resource Management at the Australian Rivers Institute, Griffith University, and a member of the International Advisory Committee of the Global Water Institute at the University of New South Wales.

Nagaraja Rao Harshadeep (Dr.) is a Global Lead for Disruptive Technology and Lead Environmental Specialist at the World Bank. In almost 25 years at the World Bank, he has led and supported several environmental, water, and natural resources operations and analytical support around the world. He has also been a Global Lead for Watersheds, currently co-leads the Bank's HydroInformatics focal area, and leads a Disruptive KIDS (Knowledge, Information & Data Services) Helpdesk. He holds a Bachelor of Technology in Civil Engineering from the Indian Institute of Technology (IIT-Madras), a Master of Science in Environmental and Resource Engineering from Syracuse, and a Ph.D. in Water Resources and Environmental Systems Engineering from Harvard University.

Editorial

Managing Water Resources in Large River Basins

William Young * and Nagaraja Rao Harshadeep

World Bank, 1818 H Street, Washington, DC 20006, USA; harsh@worldbank.org
* Correspondence: wyoung@worldbank.org

Received: 8 December 2020; Accepted: 10 December 2020; Published: 11 December 2020

The management of water resources in large rivers basins commonly involves challenges and complexities that are not found or are less common in smaller basins. Irrespective of size, issues of infrastructure construction and operation, irrigation and drainage management, water use efficiency and flood management are common. However, larger basins commonly span multiple jurisdictions and traverse diverse bio-geographies, which often give rise to greater complexity of competing interests between sectors, environments and communities. The sustainable development of large river basins commonly requires considering both consumptive water use and non-consumptive water use (e.g., inland navigation and hydropower) and tackling large-scale drought and flood management. These challenges make long-term strategic planning arguably more important than in smaller basins, but also usually more complex in both analytical, participatory and political terms. This requires navigating both "hard" issues (e.g., infrastructure, hydromet, information technologies) as well as "soft" issues (e.g., legal frameworks, policies, institutions, participation, political economy).

There is no widely accepted hard criterion for defining "large river basins" either in terms of drainage area or total discharge. Thus, herein a loose definition is adopted that includes basins exceeding 100,000 km^2 (of which there are an estimated 130 globally) as well as geographically smaller basins characterized by hydrologic complexity (e.g., extreme variability, non-stationarity, complex surface-groundwater connectivity), water management complexity (spatiotemporal supply-demand imbalance, major water infrastructure, inter-sectoral competition, pollution, high flood risk, climate change vulnerability) and/or governance complexity (cross-jurisdictional conflict, legal and regulatory complexity).

Most of the large and highly populated river basins in the world are international transboundary basins. Of the 286 international transboundary river basins in the world (spanning 151 countries and with over 40 percent of the global population), the fourteen basins with the greatest economic dependence on water are home to nearly 1.4 billion people [1]. In these and other basins, water resources management is integral to sustainable development, and in 2015, all 193 Member States of the United Nations General Assembly agreed to the 2030 Agenda for Sustainable Development. This agenda is captured in seventeen Sustainable Development Goals (SDGs), including SDG6—Ensure availability and sustainable management of water and sanitation for all. SDG6 considers water supply and sanitation services, water scarcity and water use efficiency, water quality and wastewater treatment, water ecosystems, as well as institutional aspects of water resource management and cooperation [2]. However, coordinated, efficient and effective water resources management to deliver on SDG6 (and other water-related goals) in large basins must overcome several significant challenges, as explored in the papers in this Special Issue.

This Special Issue of *Water* comprises nine papers with contributions from over fifty authors that traverse the hard and soft aspects of managing water resources in large river basins through a series of diverse case studies that showcase recent advances in technological and governance innovations for large river basin management. The papers touch on many of the great rivers of the developing world—the Ayeyarwady, Brahmaputra, Ganges, Mekong and Nile—with transboundary issues highlighted as a key challenge in many large basins.

The papers include six *Research Articles* (including the Special Issue *Feature Paper*), two *Communications* and one *Case Report*. The *Feature Paper* [3], by the guest co-editors, frames the Special Issue by discussing the distinguishing features and importance of large river basins, and introducing a conceptual framework for water security that connects water endowment, the water sector architecture (institutions and infrastructure), water sector performance (resource management, service delivery and risk mitigation) and the outcomes from how water is managed and used. Harshadeep and Young [3] then explore the application of a range of disruptive technologies to the different dimensions of water resource management, considering the disruption of data value chains, production value chains and stakeholder participation value chains. Importantly, the analysis also considers the different institutional roles in the disruption process and the important risks and barriers to adoption. Overall, Harshadeep and Young find that the increasing uptake of disruptive technologies has particular utility for large river basins and can help to "democratize" water management through improved access to data and information, but they note that an increased effort will be required to ensure equity in technology access.

Several papers in the Special Issue explore, in more detail, applications of key data technologies introduced by Harshadeep and Young. Dandridge et al. [4] present a case report for the use of Earth observations in water management in the Lower Mekong River Basin. Specifically, they describe the application of a downscaling algorithm to generate a 1 km grid for soil moisture based on the 9 km interpolated grid of soil moisture from the NASA active passive mission radiometer. Dandridge et al. demonstrate how these data can be applied directly to improve flood prediction and assessment, as well as to provide drought monitoring and agricultural productivity predictions for large river basins. Zhao et al. [5] approach the complexity of multiple competing objectives in large river basins by demonstrating the application of optimization techniques (based on cumulative probability distribution functions) to the operational management the reservoir cascade in the Heihe River in China. Zhao et al. quantify the trade-offs between competing objectives, enabling identification of "least worst" operating rules. Simonov et al. [6] approach the competing-uses challenge in the transboundary Amur River Basin by assessing the effectives of a range of analytical methods for considering the environmental impacts of both existing and proposed hydropower facilities. These methods include rapid strategic basin-scale impact assessments, assessments of flow regime and floodplain alteration, assessment of riverine habitat impacts, assessments of river fragmentation, identification of protected areas, and environmental flow assessments. The work shows that hydropower investment decisions need to be guided by a range of environmental impact assessments from the site to the basin scale, informed by the geographic distribution and connectivity of biodiversity hotspots, and integrated into a robust and participatory planning process.

Van der Vat et al. [7] present a case study of participatory modelling for the Ganges River Basin in India, that demonstrates the integration of Earth observations with in situ data within a basin modelling framework, coupled to a customized dashboard to support collaborative interactions and to guide basin planning decisions. While building on past modelling efforts for the Ganges, van der Val et al. present the first attempt to bring together analytical models for surface and groundwater hydrology, water quality, riverine ecology within an IWRM framework to explore the potential impacts of both future climate change and socio-economic development for the Ganges Basin. The collaborative modelling approach spanned central and multiple basin state governments, as well as NGOs and development partners. O'Sullivan et al. [8] focus specifically on the challenges of stakeholder engagement in large river basins. They present a new integrative framework for stakeholder engagement built around a cloud-based web application for basin modelling and planning that relies heavily on open access Earth observation datasets. They explore the utility of the platform through three case studies that explore irrigation development, transboundary water sharing, and environmental water allocations.

Foran et al. [9] explore stakeholder participation in large basins, putting technology to one side and exploring river basin decision making from collaborative governance and deliberative process standpoints. They make the case for co-production of knowledge and planning scenarios to improve the social and political legitimacy of basin planning decisions. They illustrate the approach using cases studies from the Ayeyarwady River Basin in Myanmar and the Kamala River Basin in Nepal. Despite the difficulties faced in implementing a collaborative co-production approach in these river basins, Foran et al. note that more typical bureaucratic approaches are similarly complex and yet are more contested and less likely to deliver widely accepted courses of action.

Barua et al. [10] explore the specific basin governance challenge of transboundary cooperation, focusing on options to progress cooperation in cases where political constraints preclude formal diplomacy. They explain the process and value of informal "track 2" dialogues and illustrate these with a case of a sustained four-country dialogue process in the Brahmaputra River Basin in South Asia. They demonstrate the pathways by which informal dialogue can develop a shared understanding amongst stakeholders and how this can influence basin planning and development. Gari et al. [11] investigate a formal approach to water allocation in transboundary river basins based on the principles of the United Nations' Watercourse Convention. They develop a set of potential indicators for these principles and evaluate the application of selected indicators to water allocation amongst the countries of the Nile Basin. To guide indicator selection and to assess indicator utility and acceptability, they distributed an email questionnaire to more than 200 experts from Nile basin countries and across the world. Based on questionnaire responses, they identify a subset of indicators for which there is a high level of consensus and which could provide a foundation for the application of the UN Watercourse Convention in the Nile Basin, based on the principles of equitable and reasonable use.

This Special Issue effectively highlights the significant complexities and governance challenges of managing water resources in large river basins. It highlights how disruptive technologies can help to address complex data and analytical challenges, as well as supporting effective participatory decision making over contested water resources. These aspects are illustrated by a selection of diverse case studies. Technology, however, clearly cannot solve all the governance challenges at play, and careful consideration and deployment of deliberative governance and dialogue processes are required to navigate the complexity of water politics in large river basins. This is especially the case in international transboundary river basins of high economic, environmental and political significance. Furthermore, while both hard and soft measures can deliver benefits, neither are sufficiently effective when applied alone. Effective water management in large basins requires progress in both domains and resource managers should increasingly seek to exploit synergies between technology and governance innovations.

Author Contributions: The two authors made equal contributions to this editorial. All authors have read and agreed to the published version of the manuscript.

Funding: This research received no external funding.

Acknowledgments: The authors appreciate the efforts of the *Water* editors and publication team at MDPI and the anonymous reviewers for their invaluable comments.

Conflicts of Interest: The authors declare no conflict of interest.

References

1. UNEP-DHI; UNEP. *Transboundary River Basins: Status and Trends*; United Nations Environment Programme (UNEP): Nairobi, Kenya, 2016.
2. United Nations. *Sustainable Development Goal 6 Synthesis Report on Water and Sanitation*; United Nations: New York, NY, USA, 2018.
3. Harshadeep, N.R.; Young, W. Disruptive Technologies for Improving Water Security in Large River Basins. *Water* **2020**, *12*, 2783. [CrossRef]
4. Dandridge, C.; Fang, B.; Lakshmi, V. Downscaling of SMAP Soil Moisture in the Lower Mekong River Basin. *Water* **2020**, *12*, 56. [CrossRef]
5. Zhao, M.; Huang, S.; Huang, Q.; Wang, H.; Leng, G.; Liu, S.; Wang, L. Copula-Based Research on the Multi-Objective Competition Mechanism in Cascade Reservoirs Optimal Operation. *Water* **2019**, *11*, 995. [CrossRef]

6. Simonov, E.A.; Nikitina, O.I.; Egidarev, E.G. Freshwater Ecosystems versus Hydropower Development: Environmental Assessments and Conservation Measures in the Transboundary Amur River Basin. *Water* **2019**, *11*, 1570. [CrossRef]
7. van der Vat, M.; Boderie, P.; Bons, K.C.A.; Hegnauer, M.; Hendriksen, G.; van Oorschot, M.; Ottow, B.; Roelofsen, F.; Sankhua, R.; Sinha, S.; et al. Participatory Modelling of Surface and Groundwater to Support Strategic Planning in the Ganga Basin in India. *Water* **2019**, *11*, 2443. [CrossRef]
8. O'Sullivan, J.; Pollino, C.; Taylor, P.; Sengupta, A.; Parashar, A. An Integrative Framework for Stakeholder Engagement Using the Basin Futures Platform. *Water* **2020**, *12*, 2398. [CrossRef]
9. Foran, T.; Penton, D.J.; Ketelsen, T.; Barbour, E.J.; Grigg, N.; Shrestha, M.; Lebel, L.; Ojha, H.; Almeida, A.; Lazarow, N. Planning in Democratizing River Basins: The Case for a Co-Productive Model of Decision Making. *Water* **2019**, *11*, 2480. [CrossRef]
10. Barua, A.; Deka, A.; Gulati, V.; Vij, S.; Liao, X.; Qaddumi, H.M. Re-Interpreting Cooperation in Transboundary Waters: Bringing Experiences from the Brahmaputra Basin. *Water* **2019**, *11*, 2589. [CrossRef]
11. Gari, Y.; Block, P.; Assefa, G.; Mekonnen, M.; Tilahun, S.A. Quantifying the United Nations' Watercourse Convention Indicators to Inform Equitable Transboundary River Sharing: Application to the Nile River Basin. *Water* **2020**, *12*, 2499. [CrossRef]

Publisher's Note: MDPI stays neutral with regard to jurisdictional claims in published maps and institutional affiliations.

Article

Disruptive Technologies for Improving Water Security in Large River Basins

Nagaraja Rao Harshadeep * and William Young

World Bank, 1818 H Street, Washington, DC 20433, USA; w.young@worldbank.org

* Correspondence: harsh@worldbank.org

Received: 11 May 2020; Accepted: 24 September 2020; Published: 6 October 2020

Abstract: Large river basins present significant challenges for water resource planning and management. They typically traverse a wide range of hydroclimatic regimes, are characterized by complex and variable hydrology, and span multiple jurisdictions with diverse water demands and values. They are often data-poor and in many developing economies are characterized by weak water governance. Rapid global change is seeing significant changes to the pressures on the water resources of large basins, exacerbating the challenge of sustainable water management. Diverse technologies have long supported water resource planning and development, from data collection, analytics, simulation, to decision-making, and real-time operations. In the last two decades however, a rapid increase in the range, capability, and accessibility of new technologies, coupled with large reductions in cost, mean there are increasing opportunities for emerging technologies to significantly "disrupt" traditional approaches to water resources management. In this paper, we consider the application of 'disruptive technologies' in water resources management in large river basins, through a lens of improving water security. We discuss the role of different actors and institutions for water management considering a range of emerging disruptive technologies. We consider the risks and benefits associated with the use of these technologies and discuss the barriers to their widespread adoption. We obverse a positive trend away from the reliance solely on centralized government institutions and traditional modeling for the collection and analysis of data, towards a more open and dynamic 'data and knowledge ecosystem' that draws upon data services at different levels (global to local) to support water planning and operations. We expect that technological advances and cost reductions will accelerate, fueling increased incremental adoption of new technologies in water resources planning and management. Large-basin analytics could become virtually free for users with global, regional, and national development agencies absorbing the costs of development and any subscription services for end users (e.g., irrigators) to help improve water management at user level and improve economic productivity. Collectively, these changes can help to 'democratize' water management through improved access to data and information. However, disruptive technologies can also be deployed in top-down or centralized processes, and so their use is sometimes contested or misunderstood. Increased attention therefore needs to be given to ensuring equity in technology access, and to strengthening the governance context for technology deployment. Widespread adoption of disruptive technologies will require adjustments to how water professionals are trained, increased adaptiveness in water resources planning and operations, and careful consideration of privacy and cybersecurity issues.

Keywords: disruptive technology; river basins; large basins; water security; water resources management; water governance; water data; information technology; analytics

1. Introduction

Water management is a major and growing global issue for economic development and poverty reduction [1,2]. Water is essential for food and energy security [3,4], and water-related extremes of flood and droughts have significant economic and social costs [5,6]. With increasing global population and economic development, demand for water and competition between uses and users are on the rise [7,8]. Global water consumption is estimated to have increased by 40 percent in the last four decades [8], mostly for irrigation, which represents 70 percent of total global water withdrawals [9]. While everyone depends on freshwater, the importance of groundwater is often overlooked; for example, groundwater provides drinking water for 1.5–3.0 billion people [10]. The current level of global water withdrawals is approaching a planetary boundary, which if crossed would take the Earth system outside a safe operating space for humanity [11]. As result of these pressures, an estimated 4 billion people experience severe water scarcity for at least one month of the year [12]. Local water availability constraints, rapid population growth and urbanization, inadequate infrastructure, and governance shortcomings [13] mean nearly 0.7 billion people lack access to a safely managed drinking water supply [14].

Accelerating climate change is perturbing the global water cycle [15], altering the average patterns of water availability and increasing the magnitude and frequency of water-related extremes in parts of the world. These changes, however, are uncertain and still poorly understood [15–18]. Climate change increases the uncertainty in projections of water supply and demand, and increases the uncertainty in feasibility and economic performance assessments of water infrastructure [19,20].

1.1. Large River Basins—Character and Importance

There is no widely accepted criterion for defining "large river basins", either in terms of drainage area or total discharge. Basins exceeding 100,000 km^2 in area—of which there are an estimated 130 or so globally (including 22 exceeding 1,000,000 km^2)—could reasonably be considered large. However, rather than use an arbitrary criterion such, we adopt a looser definition that also includes geographically smaller basins where water management challenges are considered 'large', because of one or more of: (i) hydrologic complexity (high flows, hydrologic variability, non-stationarity, surface-groundwater interactions, multiple water sources—rainfall-runoff, snow, glacier melt); (ii) water management complexity (large population, supply-demand imbalance, inter-sectoral competition, rapid demand growth, pollution, high flood and erosion risk, climate change vulnerability),); and (iii) administrative complexity (transboundary coordination or conflict; federate-state-local coordination or conflict; governance complexity—intersecting legal, policy, regulatory frameworks).

Aside from remote and sparsely populated basins in northern Canada and Russia, most of the geographically largest river basins in the world are also international transboundary basins. A total of 286 river international transboundary river basins have been identified (spanning 151 countries and home to more than 40 percent of the global population); 80 percent of the total area and population of the transboundary basins is associated with the largest 156 basins [21]. Larger river basins tend to have a higher economic dependence on water resources, and the 14 basins with the greater economic dependence on water are home to almost 50 percent of the population of all transboundary basins—nearly 1.4 billion people [21]. In addition to international transboundary rivers, large basins within federal countries—such as the Murray–Darling in Australia—represent complex resource management challenges [22].

As well as the economic importance and the associated social values of large river basins, these systems are critical habitat for freshwater biodiversity. Rivers, lakes, and other 'wetlands' occupy just 0.8 percent of the Earth's surface but support 6 percent of all described species including 35 percent of all vertebrates [23], especially fish. Large rivers with higher flow volumes tend to support more fish species, and tropical rivers tend to have higher levels of species richness. The highest levels of riverine fish species richness are found in the Amazon, Orinoco, Tocantins, and the Paraná in South America; the Congo, the Niger Delta, and the Ogooue in Africa; and the Yangtze, Pearl, Brahmaputra, Ganges, Mekong, Chao Phraya, Sittang, and Irrawaddy in Asia [24]. Large rivers are also especially important for freshwater megafauna with slow life-history strategies and complex habitat requirements [25]. Globally, freshwater megafauna populations declined by 88% from 1970 to 2012, with mega-fishes exhibiting the greatest global decline (−94%) [25]. These major biodiversity declines highlight the conflicts between economic development and environmental protection and conservation in large river basins.

Large river basins present particular challenges and opportunities for the use of emerging technologies in support of water resources management. Geographically, large basins typically traverse a wide range of hydroclimatic regimes, and processes that characterize basin-scale hydrological behavior take place at multiple scales. These give rise to technical challenges for the design and operation of hydrometeorological data collection systems, including the integration of ground-based and remote observations. Because they often span multiple jurisdictions—within and, or between countries—large basins have institutional complexities for coordinated data collection, sharing and analysis, as well as for decision making and for coordinated real-time operational management.

1.2. Water Security as an Objective for IWRM

Integrated water resources management (IWRM) has been vigorously promoted by the international development community as a set of principles and a best practice process for planning and managing water resources [26]. IWRM has also however, been strongly criticized from both technical standpoints (for being too vague to have real utility for practical implementation [27]) and from political economy standpoints (having been dubbed "soft coercive hegemony" [28]). We find that accepting some key principles of IWRM (e.g., water systems focus, data/analytical foundation, participatory planning) but shifting to a medium-long term 'water security' outcome focus, helps to better define desired economic, social and environmental outcomes from water, and identify specific interventions to help achieve these. 'Water security' is thus conceptualized as the relationships between the water endowment, the water sector architecture (institutions and infrastructure), water sector performance (resource management, service delivery and risk mitigation) and the outcomes from how water is managed and used (Figure 1). A recent example of the application of this conceptualization is a comprehensive water security diagnostic for Pakistan [29].

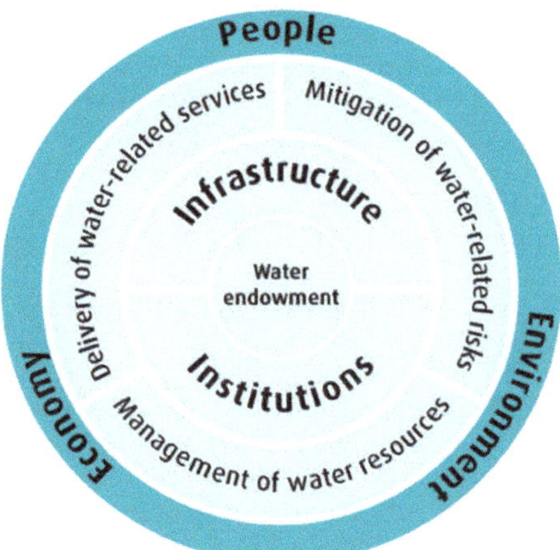

Figure 1. Conceptual framework for water security.

Water resources management is integral to sustainable development. In 2015, all 193 Member States of the United Nations General Assembly agreed to the 2030 Agenda for Sustainable Development and established 17 Sustainable Development Goals (SDGs). This is a plan to "end poverty in all its forms" and to "shift the world to a sustainable and resilient path". SDG 6—Ensure availability and sustainable management of water and sanitation for all—considers not just water supply and sanitation services, but also water scarcity and water use efficiency, water quality and wastewater treatment, water ecosystems, as well as institutional aspects of water resources management (including IWRM implementation) and cooperation [30]. Other aspects of water security (such as water-related disasters) are captured by other SDGs.

In this paper, we review the application of "disruptive technologies" in water resources management in large river basins, through this water security lens. We consider how these technologies can assist delivering better outcomes or deliver outcomes more efficiently or cost effectively. We discuss the roles of different actors and institutions, and consider risks associated with the adoption of these technologies and the barriers to widespread adoption.

1.3. Role of Technology in Water Management

Technology has multiple roles in water management, across the spheres of infrastructure design, systems planning, real-time operations. These can be considered in a matrix with the key areas of water resource management, irrigation management, water supply (and treatment) and sanitation, and environmental water management. Here, we focus on decision making—at both planning and operational time scales—for water resources management. This includes river basin planning; water allocation planning; flood and drought outlooks, forecasts and warnings; and the real-time operational management of water resources infrastructure. However, these boundaries are not tightly delineated, and many of the technologies discussed have application into other aspects of water management as well. For these selected focal areas of water resources management, we consider how 'data' is transformed into 'information' and then 'knowledge', and how these are then used in decision making for action. Along this 'value chain', we thus consider the collection, transmission, storage, management, and sharing of data. Then the ways in which data is transformed into information and thence into knowledge, and how information and knowledge are stored, managed, shared, visualized, and otherwise communicated. We consider the decision process and the roles of multiple actors in this process, and how decisions are communicated and then actioned. Beyond the 'hydro-informatics' elements of technology, there are innovative technologies for operations and stakeholder interaction. With respect to SDG 6, a brief introduction to some technology opportunities is provided by [31,32].

2. Disruptive Technologies

Disruptive technology is commonly defined as "technology that can fundamentally change not only established technologies but also the rules and business models of a given market, and often business and society overall"; the term (and concept) was first introduced in the mid-1990s by Harvard Business School scholars in the context of business innovation [33]. Disruptive technologies are now showing much promise in every field of development [34]. The evolution of these technologies is accelerating and disrupting traditional approaches to water planning and management. The key relevant technologies are summarized in Table 1 with implications on their application to water resource management in large river basins.

2.1. Technology Evolution

Technological evolution has influenced the use of water resources for millennia. However, in recent years, there has been an acceleration in the development of new tools and technologies of relevance for water resources management (Figure 2). Many factors, however, affect the adoption of emerging technologies, including institutional capacity, the enabling policy and institutional environment, resource availability, competing priorities, access to global good practice, intellectual property, and the agility of governments, the private sector, academia, and other actors. These factors have meant heterogeneous but overall slow uptake and diffusion of new technologies. Some technologies that were deployed in the developed world more than a century ago, are only now being adopted across the developing world. However, much of the developing world has a 'last mover' advantage, with the potential to leapfrog old ways and adopt new approaches more rapidly than the developmental paradigm allowed in the developed world. While there have been many challenges in leapfrogging in some areas, including for environmental sustainability and indeed for river basin management [35], information and communication technology leapfrogging is beginning to show real impact in spite of "tech transfer" and "absorptive capacity" issues [36], including through rapid adoption of mobile technologies and leveraging the large data sets generated by some developing countries [37].

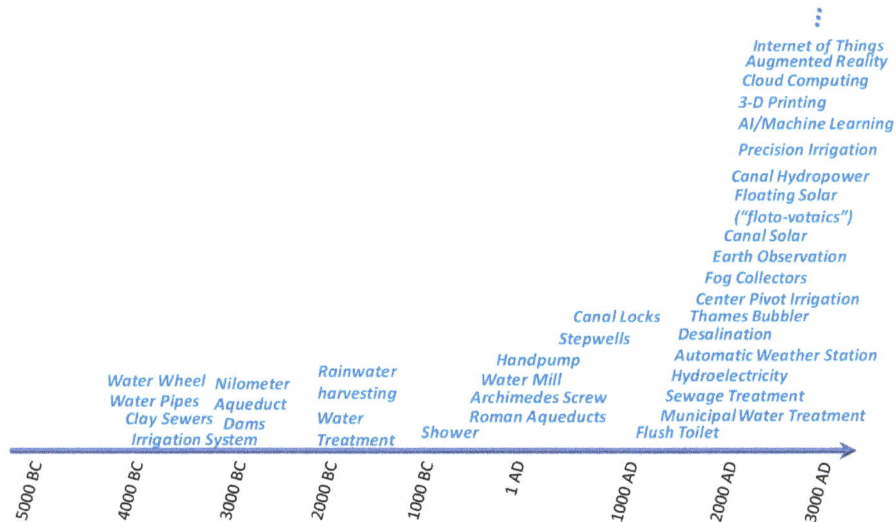

Figure 2. Timeline of water technology evolution.

Here, we classify disruptive technologies according to where they can be most disruptive: (i) decision-making, (ii) operations, and (iii) stakeholder interaction (Figure 3).

2.2. Technology Appplications

Traditional investments in water resources in the developing world have seldom been conceived, implemented, or operated from a holistic multi-sectoral basin perspective. They often are based on old technologies and have high operational and maintenance costs that are seldom met, leading to poor service delivery exacerbated by deferred maintenance as they age. Even basic monitoring data are usually not accessible in real-time and require different 'ringfenced' legacy software that are not inter-operable.

Figure 3. Typology of 'disruptive' technologies.

One of the main ways in which modern technology is reshaping water resources planning and management is through 'disrupting' the data value chain (Figure 4). This is manifested through new inexpensive sensors for in-situ monitoring (tending towards an expansive 'internet of things'), increasingly powerful Earth observations from satellites and drones/unmanned aerial vehicles (UAVs) to provide synoptic views of topography (including high-resolution digital elevation models to identify flood-prone areas and support hydrodynamic modeling), climate, water levels, flows, snow cover, inundated areas, landcover, watershed status, and even some aspects of water quality and groundwater. Earth observations [22], with near global consistent coverage, is rapidly becoming a game-changer for synoptic observations in large basins, where the resolution of even the free resources from NASA and ESA are often adequate for useful water resources analytics. New unmanned on-water and under-water vehicles show promise; they can be outfitted with sensors and autonomous (single or swarm) capability for surveying large water bodies (e.g., for bathymetry, hydraulic safety, water quality, or fish stock assessments).

New analytical tools, increasingly cloud-based—including at the global level, assimilate available data and generate estimates of a range of critical parameters related to snowmelt estimation, water balance, water accounting (e.g., WA+, WAPOR) [38,39], scenario analysis, and forecasts to create 'digital twins' of basins to facilitate analyses. These enable access to curated archives and real-time estimates of the water status for any basin anywhere in the world to support both strategic planning and tactical operations through data visualization, early alerts/warnings, and the development of interactive packaging for data, analytics, and knowledge. Examples include interactive portals, mobile phone applications, and dynamic e-books. These support decisions at all levels, from simple scoping of water resources development, to detailed planning with stakeholder involvement/outreach, as well as real-time operations. Additional systems related to data/text mining, social media integration, advanced cloud-based modeling, machine learning/AI, or 'bots' can help bring in an additional automation and integrated perspectives to support decision-making.

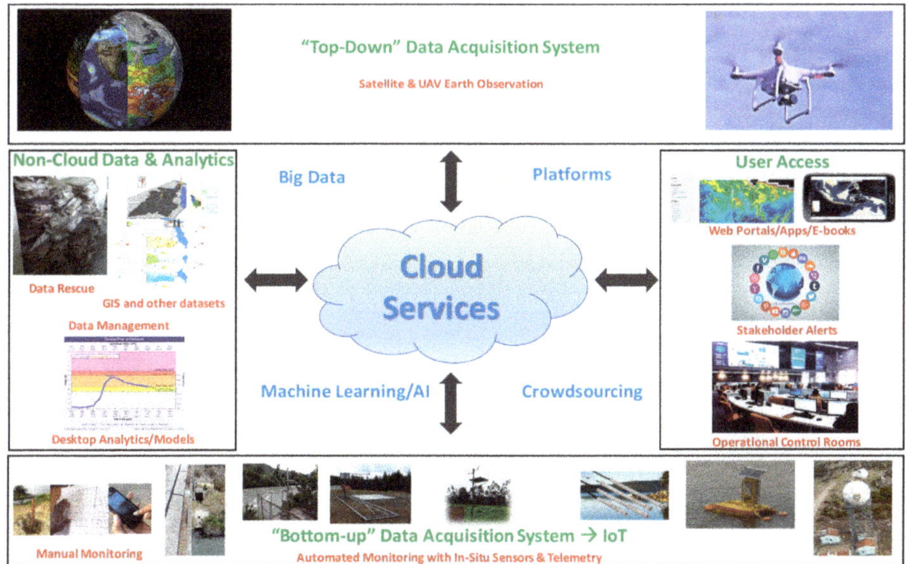

Figure 4. Modernizing the data value chain (Data→Information→Knowledge→Decision Support).

These technologies are helping water managers reimagine the way information-based decisions can be made for smart water resources planning and management and are allowing development of integrated basin/aquifer plans based on both analytical and stakeholder approaches. Technologies have made possible new approaches to use and conserve water and administer usage caps (e.g., using satellite-derived actual evapotranspiration estimates), adopting a systems perspective to improve agricultural water productivity, benchmark systems, and incentivize sustainability. New in-situ and Earth observation monitoring and analytics allow for development and customization of tools for water planning, allocation, and coordinated water infrastructure operations in an integrated multi-sectoral systems perspective. Water infrastructure can now be operated in a more coordinated systems context for multiple objectives ranging from service delivery to climate resilience. Continuous innovation, piloting, and learning from global good practices enables quick scaling-up of new technologies and enables more adaptive management.

3D printing, robotics, automated transport, advanced materials, nanotech, biotech, and cleantech are supporting new operational systems that represent a paradigm shifts away from traditional approaches. Examples include irrigation systems that improve water productivity and field-level water use efficiency (especially when combined with policies such as limits on water abstraction); 3D printed monitoring stations (e.g., 3D-PAWS [40]) that reduce costs for monitoring weather and water levels; and ultrasonic control systems (successfully piloted in Lake Quaroun in Lebanon) that can mitigate algae-related water quality problems.

Platforms are emerging to enable people to work together in new ways in the sharing economy, including fintech, crowdsourcing, crowdfunding, block-chain enabled supply chains, asset sharing systems, Digital ID enabled e-governance, and open learning platforms. Many of these platforms have application in water management in large basins, including online/mobile platforms to support learning or interactions among remote or disperse water user associations, and platforms to help farmers access global marketplaces online, with feedbacks into irrigation water requirements.

2.3. Implications for Large Basins

The implications of these new technologies, with a focus on digital technologies, for water resources planning and management in large basins, are summarized in Table 1. Major changes in water management around the world are likely in very short timeframes. Ref [41] Many of the initial impacts will come from the plummeting costs of sensors, mobile devices and connectivity, cloud services (including to process increasingly powerful earth observation and other big data), interoperability due to online data standards and protocols, and increasing digital literacy.

Water data could be used primarily for water assessments, evaluations, operations, foresight, design, accountability, and education [42]. Many of these are useful at different scales—from monitoring progress towards the SDG-6 global indicators to helping design a culvert.

Many countries are modernizing their water resources institutions and developing national water resource information systems and analytics to support basin planning and disaster management. Some are also strengthening ties among government, academic, civil society organizations (CSO), and private sector institutions to develop a broad stakeholder base for this transformation. Modern water information systems require integration of data (from global, regional, national, to local sources), data quality management, conversion to interoperable data services, and development of interactive dashboards to help access and visualize data services and associated analytics in appropriate formats to support decisions.

Large basins will especially benefit from these changes given both the challenges (e.g., the need to integrate data across large areas; multiple stakeholders wishing to inform coordinated decisions; large water infrastructure investments) and the opportunities of large basins (e.g., application of free Earth observations in the 10–250 m resolution range; the ability to deliver reach large numbers of remote beneficiaries with valuable data services).

Institutions such as the Mekong River Commission and the various Nile Basin Initiative centers have demonstrated the utility of modern data and analytics in basin planning and hydro-meteorological data integration. Other large basins (e.g., the Congo, Ganges-Brahmaputra-Meghna) are in the nascent stages of this journey given capacity constraints and transboundary cooperation challenges.

Many countries are modernizing their water information systems taking advantage of new technologies. The United States and Australia are improving their already well-established systems facilitated by strong national institutions. Europe is increasingly building on its regional institutions (e.g., European Centre for Medium-Range Weather Forecasts, European Organisation for the Exploitation of Meteorological Satellites, Joint Research Centre of the European Commission) to help countries access better data and analytics. China is utilizing evapotranspiration estimates derived from Earth observations to improve irrigation management [43,44] and India is enhancing its national water resources information system [45].

Estimates based on satellite products or global models are increasingly found to be comparable with those based on in-situ observations [46,47]. These techniques, especially when enhanced by a new generation of artificial intelligence (AI)/machine-learning (ML) algorithms and global models, can revolutionize water resources management even in data-poor environments. When accessed through customized interactive dashboards, this information can be especially useful for estimating parts of the water balance, estimating flooding areas, making customized weather/hydrologic/inundation forecasts, managing large water demands (e.g., agriculture) and system losses, while enhancing and benchmarking water productivity [48,49].

Table 1. Summary of disruptive technologies relevant to water management in large basins.

Technology	Description	Implications for Large Basins	References
In-situ Sensors and Internet of Things (IoT)	Sensor is a generic term for devices capable of sensing external stimuli (e.g., force, flow, acceleration, light, sound, vibration, humidity, temperature, pressure) and act upon those readings (e.g., recording, reporting, reacting). IoT refers to a set of physical objects with embedded ubiquitous sensors connected to networks (e.g., telemetry) and interfacing with analytics/applications that support real-time management.	Inexpensive in-situ real-time monitoring networks (e.g., for snow, flows, soil moisture, groundwater, water quality) especially with an internet-of-things approach supported by effective telemetry (GSM, satellite, radio, blue-tooth, broadband, etc.) for natural (e.g., streams, rivers, lakes, coasts) and man-made systems (e.g., canals, pipes)	[50,51]
Earth Observation and Geospatial	Remote sensing (acquisition and processing of information without making contact) using satellites, drones/UAVs and other aircraft. Analytics using modern GIS and Remote Sensing Software and Services.	Satellite data products for weather, land cover, water levels, evapotranspiration, flow, and groundwater change for large basins. Heliborne surveys to explore geological structure; drone surveys of streambanks; LiDAR surveys for flood-prone areas; drone/UAV surveys. Geospatial data processing tools (from desktop to online systems) support applications from data visualization to complex modeling.	[52–54]
Cloud Services	A cloud service is any service made available to users on demand via the internet from cloud-computing servers. Cloud services are designed to provide easy, scalable access to applications, resources and services, and are fully managed by a cloud services provider. As cloud services become more ubiquitous, cheaper, and more secure, they offer more opportunities for combining data in new ways and making them accessible on multiple devices in any setting.	'Big data' analytics especially when supported by new data science advances in scripting, online analytics, modelling, and visualizations. These services can be free, or subscription based (e.g., Google Earth Engine, Open Data Cube, etc.)	[52,55–57]
Open Data/Analytics and Standards	Open standards for data/data and analytics services (e.g., Open Geospatial Consortium (OGC), Open APIs) are helping make data and analytics more accessible in online contexts. The concept of AI tries to mimic human intelligence in machines. ML enables computer programs to become self-learning through data mining and supervised, unsupervised, or hybrid training systems. Many countries are encouraging policies to promote more data accessibility in the public domain. Blockchain could help use a distributed ledger technology that can help make transactions relatively tamper-proof.	Open standards help make systems interoperable. Some of these (e.g., WaterML from OGC) have a lot of potential if their use becomes widespread. Open data can be particularly useful for weather, streamflows, soil moisture, groundwater, water quality, crop yields, etc. to facilitate decision support and training for machine learning/AI systems. Models (especially in the public domain) for water balance or water systems analysis are slowly moving to cloud platforms and could disrupt traditional desktop modeling when better established. ML/AI systems can help integrate disparate information, power automation, develop chatbots, and aid language translation. Blockchain could improve transparency and reliability of data reporting, smart contracts, and water administration.	[58–63]
User Interfaces (portals, mobile Apps, augmented/virtual reality)	Rapid advances are being made in providing access to data, analytics, and knowledge to support learning and decision-support for end-users, as well as new ways to crowdsource user observations, surveys, and other inputs.	Interfaces for modern operational control rooms/water centers, smartphones, tablets, computers, augmented/virtual/mixed reality devices to access, visualize, and analyze basin data from in-situ sensors, Earth observations, model results, etc.	[64,65]
Stakeholder Interaction	Tremendous improvements are being made in connectivity (mobile voice and data access; broadband), and in e-government/private sector online services.	Improved connectivity through mobile devices and online services can usher in a new paradigm for stakeholders to work together and access global good practice	[66–68]

3. Institutional Roles

Traditionally, water resources institutions have faced the challenge of inefficient workflows, low capacity, poor coordination across sectors and governance levels, poor integration with other kinds of institutions (e.g., academia, private sector), limited transparency, and inadequate alacrity in learning from global good practices. New technologies are enabling improvements in information interoperability [63] and institutional infrastructure that can be developed at reasonable cost (e.g., ecosystems of computers, tablets, and smartphones and related Apps; shared audio/videoconferencing, shared communication, and touchscreen access resources; dynamic physical-computer modeling approaches [e.g., [69]). These can be co-located in clusters such as water centers that allow for co-location of representatives from related organizations that need to work together (e.g., the National Water Center in the United States) to develop and use shared products (e.g., the National Water Model [70]). These approaches could be adopted in many countries and transboundary basins where shared personnel and tools could 'disrupt' traditional 'siloed' approaches.

New technologies are also fundamentally changing the notion of capacity-building, with services, automation, and interfaces reducing the need for laborious and time-consuming issues related to access restrictions, digitization, formats, fragmented desktop analysis, and dissemination on a case-by-case basis, all with only a few people having access to even view the data and products. New more automated systems enable wider access and a different kind of capacity development, avoiding the need for reinventing similar systems at great cost and with limited new functionality. These systems can increase levels of collaboration between agencies, through shared data, analytics, and visualization services. All stakeholders can leverage the learning and collaboration systems supported by the internet and high-speed connectivity, in order to learn rapidly from and contribute to global good practices. New technologies can help redesign stakeholder consultation, climate hazard insurance, payments for ecosystem services, by connecting stakeholders and by accessing new data analytics. Different institutions have different roles in this evolving world (Table 2).

Table 2. Institutional roles in the disruption process.

Institution	Potential Role in the Disruption Process
Governments	Enabling policy environment for innovation (e.g., open data policies, incentivizing collaboration and innovation, building/facilitating the backbone cyberinfrastructure, improving internal and external collaboration and shared vision, creating internship/visiting expert programs, open transparent procurement and learning expos to facilitate innovation, shared vision and collaborative decision-making) as well as managing the downside risks (e.g., obsolete jobs, privacy, cybersecurity). The role would be customized to the level of government institution (from national to provincial to local) considering the opposite implications of the principles of subsidiarity and economies of scale.
Academia	Improving research and data/tools/literature in the public domain, educate existing water professionals and a new generation of water professionals on the potential for new technologies, collaborative research and internship programs, contributions in hackathons and other competitions.
Private Sector	Develop innovative approaches that respond to challenges faced by various stakeholders, showcase new approaches, explore opportunities to demonstrate proof-of-concept.
Regional and Global Institutions	Facilitate access to finance, knowledge of regional and global good practices, learning and collaboration (e.g., for transboundary basin organizations, multilateral or bilateral development organizations, large CSO, partnerships, etc.) related to the use of new technologies and sharing lessons from implementation experience.
Community	Improve awareness of emerging disruptive technologies and role of the public in highlighting opportunities and concerns and demanding and using open data for action and social media. Increase and improve public involvement through CSO facilitation, citizen science approaches, and crowdsourcing/crowdfunding innovations.

Water sector institutions have both management and governance roles [71] and both these roles can be enhanced through technology. For example, in the case of groundwater management in large areas, improved monitoring using in-situ sensors (e.g., for extraction, use, quality, recharge), Earth observations (e.g., for evapotranspiration estimates and gravity-based water equivalent changes) and improved models, could improve resource management through regulating groundwater pumping (volumes and timing) or determining bore spacings. Modern communication including for automated collection of fines, could improve governance to promote resource management goals of equity, efficiency, conservation, and sustainability. New technologies could be particularly useful in transboundary waters contexts, with multiple options now available for estimating resource extent and condition, and other key water resources variables.

A key constraint in reaching the potential the various technologies is data availability. This could be overcome by wider government adoption of an open data approach. For example, the California Open and Transparent Water Data Act requires the state Department of Water Resources to create, operate, and maintain an open-access state-wide integrated water data platform [72].

4. Benefits, Risks, and Barriers to Adoption

The technologies discussed herein offer important technical and governance benefits compared to traditional approaches. They can increase robustness in decision making, as decisions are more likely to be based on more complete information (e.g., from weather and other apps, portals, decision support systems). They can increase the timeliness and accuracy of real-time and near real-time decisions through greater use of automation and rapid and reliable communications. They can reduce the costs of basin planning and management (e.g., lower traditional hydromet monitoring costs, reduced redundancy, and increased economies of scale from online services). They benefit end-users through better information and decision-support and enhanced mechanisms to connect stakeholders and global good practice (e.g., through social media and packaged curated content). They provide enhance trust and cooperation across sectors and regions (including transboundary) and can support more democratic decision-making through open and equal access to data and information.

While disruptive technologies (even while often benefitting from centralized platforms and standards) are encouraging a move away from top-down centralized decision processes, their adoption does not guarantee this positive shift. There are many instances of the deployment of disruptive technologies in top-down or centralized decision making without adequate stakeholder consultation. Understanding the governance context for disruptive technology deployment is therefore critical, as is explicit consideration of how this context determines whether new technologies enhance or hinder processes of stakeholder participation and empowerment [73,74].

In addition to these potential negative consequences, there are significant barriers to widespread and rapid uptake of these technologies. Adoption requires a considerable range of new technical skills, many of which are not standard in university water management curricula. In addition to awareness-building, there is a need to improve the sector skill-base through training, recruiting/ insourcing appropriate cutting-edge technology skills, and building partnerships. As with any new technology, the need to invest in new infrastructure has budget implications, and while costs for many of the technologies are rapidly reducing, governments may be reluctant to invest in what may be perceived as non-standard equipment. As these technologies are evolving rapidly, there will be pressure to update and upgrade more frequently than in the past. With the accelerated risk of obsolescence, it will be important to move to new adaptable cloud-based approaches that allow rapid upgrading of systems. This highlights the need for changed mindsets to help water resource managers and decision makers step out of the 'comfort zone' to recognize this new world of rapidly evolving technologies.

As well as barriers to adoption, there are some significant risks. Effective management of privacy and cybersecurity risks requires good institutional policies, frameworks, and systems [71]. There are implications for professional employment, since as with any technological change, large numbers of employees will increasingly find there is diminishing need for traditional, manual jobs as these become automated (e.g., gauge readers, analysts, desktop modelers, translators, etc.). Employers will need to recognize these trends and institute retraining or retrenchment/skill upgrading/replacement programs for effective workforce management. There is a risk that the digital divide will become greater—with many countries, and communities within countries, unable to access the disruptions that seem to be changing life for the better in other places and for other people. This will require increased emphasis on low-cost or free open public-domain systems and the ability to create and use more global platforms.

5. Conclusions and Forward Look

A new world of innovative technologies has the potential to 'disrupt' traditional approaches to water resources management in large basins. The widespread operationalization of the fuzzy concept of IWRM is now within reach, with new ways to strengthen the information, institution, and investment foundations of IWRM.

Looking ahead, there are two mutually reinforcing aspects that will help make the rate of technological adoption exponential. First, technology is evolving at a blistering pace, dropping the costs for every process and enabling actions that were not even considered in the realm of possibility a few years ago. Second, the incremental adoption of some of these technological options in water resources planning and management are generating lessons that can inspire others to do even better as adoption spreads.

These changes are likely to lead to a new way of reconsidering data and analytical sovereignty as data and analytics (e.g., for droughts, floods, basin scenario planning) become increasingly global, fueled by machine learning that builds on opening up of data access for training. Water withdrawal and net consumption will be closely tracked and monitored (with a combination of in-situ sensors and Earth observation) to improve systems management and benchmarking. Large-basin analytics could become virtually free for users with global, regional, and national development agencies absorbing the costs of development and any subscription services. Services for end users (e.g., farmers) could also become free or low-cost services and help improve water management at the user level and deliver improved productivity.

Collectively, these changes can help to 'democratize' water management through improved access to data and information, but increased attention will need to be given to equity in technology access. Disruptive technologies will require adjustments to how water professionals are trained, an increasing adaptiveness in water resources planning and operations, and careful consideration of privacy and cybersecurity issues. Especially as the world struggles with the ongoing COVID-19 pandemic experience, there is an increasing appreciation of the use of such new technologies to help conceive, remotely monitor, and manage water resource systems and related investments. Strong leadership to create an enabling environment to improve awareness and skills related to new technologies to realize the promised benefits and effectively manage risks is essential to facilitate this modernized approach to planning and managing large basins.

Author Contributions: Conceptualization, N.R.H. and W.Y.; Methodology, N.R.H. and W.Y.; Writing—Original Draft Preparation, N.R.H.; Writing—Review and Editing, W.Y. All authors have read and agreed to the published version of the manuscript.

Funding: This research received no external funding.

Acknowledgments: We thank three anonymous reviewers for improvements to the manuscript.

Conflicts of Interest: The authors declare no conflict of interest.

References

1. Oki, T.; Kanae, S. Global hydrological cycles and world water resources. *Science* **2006**, *313*, 1068–1072. [CrossRef] [PubMed]
2. Postel, S.L.; Daily, G.C.; Ehrlich, P.R. Human appropriation of renewable fresh water. *Science* **1996**, *271*, 785–788. [CrossRef]

3. Lotze-Campen, H.; Müller, C.; Bondeau, A.; Rost, S.; Popp, A.; Lucht, W. Global food demand, productivity growth, and the scarcity of land and water resources: A spatially explicit mathematical programming approach. *Agric. Econ.* **2008**, *39*, 325–338. [CrossRef]
4. Wan, L.; Wang, C.; Cai, W. Impacts on water consumption of power sector in major emitting economies under INDC and longer-term mitigation scenarios: An input-output based hybrid approach. *Appl. Energy* **2016**, *184*, 26–39. [CrossRef]
5. Centre for Research on the Epidemiology of Disasters. EM-DAT—The International Disaster Database. Available online: http://www.emdat.be/database (accessed on 26 May 2020).
6. Carolwicz, M. Natural hazards need not lead to natural disasters. *EOS* **1996**, *77*, 149–153. [CrossRef]
7. Distefano, T.; Kelly, S. Are we in deep water? Water scarcity and its limits to economic growth. *Ecol. Econ.* **2017**, *142*, 130–147. [CrossRef]
8. Qin, Y.; Mueller, N.D.; Siebert, S.; Jackson, R.B.; AghaKouchak, A.; Zimmerman, J.B.; Tong, D.; Hong, C.; Davis, S.J. Flexibility and intensity of global water use. *Nat. Sust.* **2019**, *2*, 515–523. [CrossRef]
9. WWAP (World Water Assessment Program). *The United Nations World Water Development Report 3: Water in a Changing World*; UNESCO: Paris, France; Earthscan: London, UK, 2009.
10. MEA (Millennium Ecosystem Assessment). *Ecosystems and Human Well-Being: Wetlands and Water Synthesis*; World Resources Institute: Washington, DC, USA, 2005.
11. Steffen, W.; Richardson, K.; Rockström, J.; Cornell, S.E.; Fetzer, I.; Bennett, E.; Biggs, R.; Carpenter, S.R.; De Vries, W.; De Wit, C.A.; et al. Planetary boundaries: Guiding human development on a changing planet. *Science* **2015**, *347*. [CrossRef]
12. Mekonnen, M.M.; Hoekstra, A.Y. Four billion people facing severe water scarcity. *Sci. Adv.* **2016**, *2*, e1500323. [CrossRef]
13. IEG (Independent Evaluation Group). *A Thirst for Change. World Bank Publications*; No. 29345; World Bank: Washington, DC, USA, 2017.
14. WHO and UNICEF (World Health Organization and UNICEF Joint Monitoring Programme (JMP). Progress on Drinking Water and Sanitation, 2015 Update and MDG Assessment. Available online: http://water.org/water-crisis/water-sanitation-facts/ (accessed on 10 May 2020).
15. IPCC (Intergovernmental Panel on Climate Change). *Climate Change 2014: Impacts, Adaptation, and Vulnerability. Part A: Global and Sectoral Aspects. Contribution of Working Group II to the 5th Assessment Report of the IPCC*; Field, C.B., Barros, V.R., Dokken, D.J., Mach, K.J., Mastrandrea, M.D., Bilir, T.E., Chatterjee, M., Ebi, K.L., Estrada, Y.O., Genova, R.C., et al., Eds.; Cambridge University Press: Cambridge, UK; New York, NY, USA, 2014.
16. Wasko, C.; Sharma, A. Global assessment of flood and storm extremes with increased temperatures. *Sci. Rep.* **2017**, *7*, 7945. [CrossRef]
17. Sharma, A.; Wasko, C.; Lettenmaier, D.P. If precipitation extremes are increasing, why aren't floods? *Water Resour. Res.* **2018**. [CrossRef]
18. Dai, A. Increasing drought under global warming in observations and models. *Nat. Clim. Chang.* **2012**, *3*, 52–58. [CrossRef]
19. World Bank. *The Critical Face of Climate Change—Water*; World Bank: Washington, DC, USA, 2016.
20. Ray, P.A.; Brown, C.M. *Confronting Climate Uncertainty in Water Resources Planning and Project Design: The Decision Tree Framework*; World Bank: Washington, DC, USA, 2015.
21. UNEP-DHI; UNEP. *Transboundary River Basins: Status and Trends*; United Nations Environment Programme (UNEP): Nairobi, Kenya, 2016.
22. García, L.; Rodríguez, J.; Wijnen, M.; Pakulski, I. *Earth Observation for Water Resources Management: Current Use and Future Opportunities for the Water Sector*; World Bank: Washington, DC, USA, 2016.
23. Garrick, D.; Anderson, G.; Connell, D.; Pittock, J. Federal rivers: A critical overview of water governance challenges in federal systems. In *Federal Rivers: Managing Water in Multi-Layered Political Systems*; Garrick, D., Anderson, G., Connell, D., Pittock, J., Eds.; Edward Elgar Publishing: Cheltenham, UK; Northampton, MA, USA, 2014; pp. 3–19.
24. Dudgeon, D.; Arthington, A.H.; Gessner, M.O.; Kawabata, Z.-I.; Knowler, D.J.; Lévêque, C.; Naiman, R.J.; Prieur-Richard, A.-H.; Soto, D.; Stiassny, M.L.J.; et al. Freshwater biodiversity: Importance, threats, status and conservation challenges. *Biol. Rev.* **2016**, *81*, 163–182. [CrossRef] [PubMed]

25. Revenga, C.; Tyrrell, T. Major river basins of the world. In *The Wetland Book*; Finlayson, C., Milton, G., Prentice, R., Davidson, N., Eds.; Springer: Dordrecht, The Nethrelands, 2016.
26. He, F.; Zarfl, C.; Bremerich, V.; David, J.N.W.; Hogan, Z.; Kalinkat, G.; Tockner, K.; Jähnig, C.J. The global decline of freshwater megafauna. *Glob. Chang. Biol.* **2019**, 1–10. [CrossRef] [PubMed]
27. Rahaman, M.M.; Varis, O. Integrated water resources management: Evolution, prospects and future challenges. *Sustai. Sci. Pract. Policy* **2005**, *1*, 15–21. [CrossRef]
28. Biswas, A.K. Integrated water resources management: Is it working? *Int. J. Water Resour. Dev.* **2008**, *24*, 5–22. [CrossRef]
29. Allouche, J. The birth and spread of IWRM—A case study of global policy diffusion and translation. *Water Altern.* **2016**, *9*, 412–433.
30. Young, W.J.; Anwar, A.; Bhatti, T.; Borgomeo, E.; Davies, S.; Garthwaite, W.R., III; Gilmont, E.M.; Leb, C.; Lytton, L.; Makin, I.; et al. *Pakistan: Getting More from Water. Water Security Diagnostic*; World Bank: Washington, DC, USA, 2019.
31. United Nations. *Sustainable Development Goal 6 Synthesis Report on Water and Sanitation*; United Nations: New York, NY, USA, 2018.
32. Borden, C.; Swanson, D.; Young, W. Monitoring SDG progress for water and sanitation: Challenges and opportunities. *IAHR Hydrolink* **2017**, *3*, 88–91.
33. Bower, J.L.; Christensen, C.M. Disruptive Technologies: Catching the Wave. *Harv. Bus. Rev.* **1995**, *73*, 43–53.
34. E-book Primer on Disruptive Technology, World Bank. Available online: www.appsolutelydigital.com/dt/ (accessed on 10 May 2020).
35. Shah, T.; Makin, I.; Sakthivadivel, R. Limits to leapfrogging: Issues in transposing successful river basin management institutions in the developing world. In *Intersectoral Management of River Basins, Proceedings of an International Workshop on Integrated Water Management in Water-Stressed River Basins in Developing Countries: Strategies for Poverty Alleviation and Agricultural Growth, Loskop Dam, South Africa, 16–21 October 2000*; Abernethy, C.L., Ed.; International Water Management Institute: Colombo, Sri Lanka, 2001; pp. 89–114.
36. Steinmueller, W.E. ICTs and the possibilities for leapfrogging by developing countries. *Int. Labour Rev.* **2020**, *140*, 193–210. [CrossRef]
37. Ciuriak, D.; Ptashkina, M. A Global South Strategy to Leverage the Digital Transformation for Development. World Trade Organization, Enhanced Integrated Framework (EIF) Aid for Trade Series. 2019. Available online: https://ssrn.com/abstract=3405330 (accessed on 10 May 2020).
38. Karimi, P.; Bastiaanssen, W.G.M.; Molden, D. Water Accounting Plus (WA+)—A water accounting procedure for complex river basins based on satellite measurements. *Hydrol. Earth Syst. Sci.* **2013**, *17*, 2459–2472. [CrossRef]
39. FAO. Available online: https://wapor.apps.fao.org/home/WAPOR_2/2 (accessed on 10 May 2020).
40. 3D-PAWS. 3D Printed Automatic Weather Station Initiative. Available online: https://www.icdp.ucar.edu/core-programs/3dpaws/ (accessed on 10 September 2020).
41. World WaterNet. Ultra-Sonic Algae Control Devices and Algae monitoring stations Installed in Lake Qaraoun, Lebanon. 2018. Available online: www.wereldwaternet.nl/en/latest-news/2018/august/ultra-sonic-algae-control-devices-installed-in-lake-qaraoun/ (accessed on 10 September 2020).
42. Bureau of Meteorology. *Good Practice Guidelines for Water Data Management Policy: World Water Data Initiative*; Bureau of Meteorology: Melbourne, Australia, 2017.
43. Wu, B.; Jiang, L.; Yana, N.; Perry, C.; Zeng, H. Basin-wide evapotranspiration management: Concept and practical application in Hai Basin, China. *Agric. Water Manag.* **2013**, *145*, 145–153. [CrossRef]
44. Wu, B.; Meng, J.; Li, Q.; Yan, N.; Du, X.; Zhang, M. Remote sensing-based global crop monitoring: Experiences with China's Cropwatch system. *Int. J. Digit. Earth* **2014**, *7*, 113–137. [CrossRef]
45. Central Water Commission. Water Resources Information System. 2019. Available online: Cwc.gov.in/water-resources-information-system-wris (accessed on 10 September 2020).
46. Le, M.-H.; Lakshmi, V.; Bolten, J.; Du Bui, D. Adequacy of satellite-derived precipitation estimate for hydrological modelling in Vietnam basins. *J. Hydrol.* **2020**, *586*. [CrossRef]
47. Dandridge, C.; Fang, B.; Lakshmi, V. Downscaling of SMAP soil moisture in the lower Mekong river basin. *Water* **2020**, *12*, 56. [CrossRef]
48. Harshadeep, N. *Innovations for Sustainable Planning and Management of Watersheds, World Meteorological Organization*; World Meteorological Organization: Geneva, Switzerland, 2018; Volume 67.

49. Harshadeep, N. Spatial agent: Reimagining data & analytics in an increasingly online world, sector insight. *Photogramm. Eng. Remote Sens.* **2019**, *85*, 707–711.
50. Robles, T.; Alcarria, R.; Martin, D.; Morales, A.; Navarro, M.; Calero, R.; Iglesias, S.; López, M. An internet of things-based model for smart water management. In Proceedings of the 28th International Conference on Advanced Information Networking and Applications Workshops, Victoria, BC, Canada, 13–16 May 2014; pp. 821–826. [CrossRef]
51. Perumal, T.; Sulaiman, M.N.; Leong, C.Y. Internet of Things (IoT) enabled water monitoring system. In Proceedings of the IEEE 4th Global Conference on Consumer Electronics (GCCE), Osaka, Japan, 27–30 October 2015; pp. 86–87. [CrossRef]
52. Bucur, A.; Wagner, W.; Elefante, S.; Naeimi, V.; Briese, C. Development of an Earth observation cloud platform in support to water resources monitoring. In *Earth Observation Open Science and Innovation. ISSI Scientific Report Series*; Mathieu, P.P., Aubrecht, C., Eds.; Springer: Berlin, Germany, 2018; Volume 15. [CrossRef]
53. Lawford, R.; Strauch, A.; Toll, D.; Fekete, B.; Cripe, D. Earth observations for global water security. *Curr. Opin. Environ. Sustain.* **2013**, *5*, 633–643. [CrossRef]
54. McCabe, M.F.; Rodell, M.; Alsdorf, D.E.; Miralles, D.G.; Uijlenhoet, R.; Wagner, W.; Lucieer, A.; Houborg, R.; Verhoest, N.E.C.; Franz, T.E.; et al. The future of Earth Observation in hydrology. *Hydrol. Earth Syst. Sci.* **2017**, *21*, 3879–3914. [CrossRef]
55. Nie, X.; Fan, T.; Wang, B.; Li, Z.; Shankar, A.; Manickam, A. Big data analytics and IoT in operation safety management in under water management. *Comput. Commun.* **2020**, *154*, 188–196. [CrossRef]
56. Lewis, A.; Oliver, S.; Lymburner, L.; Evans, B.; Wyborn, L.; Mueller, N.; Raevski, G.; Hooke, J.; Woodcock, R.; Sixsmith, J.; et al. The Australian Geoscience Data Cube—Foundations and lessons learned. *Remote Sens. Environ.* **2017**, *202*, 276–292. [CrossRef]
57. Gorelick, N.; Hancher, M.; Dixon, M.; Ilyushchenko, S.; Thau, D.; Moore, R. Google Earth Engine: Planetary-scale geospatial analysis for everyone. *Remote Sens. Environ.* **2017**, *202*, 18–27. [CrossRef]
58. Dogo, E.M.; Salami, A.F.; Nwulu, N.I.; Aigbavboa, C.O. Blockchain and internet of things-based technologies for intelligent water management system. In *Artificial Intelligence in IoT. 2019. Transactions on Computational Science and Computational Intelligence*; Al-Turjman, F., Ed.; Springer: Berlin, Germany, 2019. [CrossRef]
59. Chohan, U.W. Blockchain and Environmental Sustainability: Case of IBM's Blockchain Water Management. Notes on the 21st Century (CBRI). 2019. Available online: https://ssrn.com/abstract=3334154 (accessed on 10 May 2020).
60. Blodgett, D.; Dornblut, I. (Eds.) *OGC WaterML 2: Surface Hydrology Features (HY_Features)—Conceptual Model, Version 1.0*; Open Geospatial Consortium: Wayland, MA, USA, 2018. [CrossRef]
61. Khattar, R.; Ames, D.P. A web services-based water data sharing approach using OGC Standards. *Open Water J.* **2020**, *6*, 2.
62. Reichstein, M.; Camps-Valls, G.; Stevens, B.; Jung, M.; Denzler, J.; Carvalhais, N.; Prahbat. Deep learning and process understanding for data-driven Earth system science. *Nature* **2019**, *566*, 195–204. [CrossRef] [PubMed]
63. World Economic Forum. Harnessing the fourth industrial revolution for water. In *Fourth Industrial Revolution for the Earth Series*; World Economic Forum: Cologny, Switzerland, 2018; p. 25.
64. Wang, J.; Wang, T.; Zhang, J.; Tan, H.; He, L.; Cheng, J. Study on the construction and application of 3D visualization platform for the Yellow River Basin. In *Frontiers of WWW Research and Development. Lecture Notes in Computer Science, 3841*; Zhou, X., Li, J., Shen, H.T., Kitsuregawa, M., Zhang, Y., Eds.; Springer: Berlin/Heidelberg, Germany, 2006.
65. Yong, T.; Zheng, Y.; Zheng, C. Development of a visualization tool for integrated surface water–groundwater modeling. *Comput. Geosci.* **2016**, *86*, 1–14.
66. Mackay, E.B.; Wilkinson, M.E.; Macleod, C.J.A.; Beven, K.; Percy, B.J.; Macklin, M.G.; Quinn, P.F.; Stutter, M.; Haygarth, P.M. Digital catchment observatories: A platform for engagement and knowledge exchange between catchment scientists, policy makers, and local communities. *Water Resour. Res.* **2015**, *51*, 4815–4822. [CrossRef]
67. Medema, W.; Furber, A.; Adamowski, A.; Zhou, Q.; Mayer, I. Exploring the potential impact of serious games on social learning and stakeholder collaborations for transboundary watershed management of the St. Lawrence River Basin. *Water* **2016**, *8*, 175. [CrossRef]

68. Farjad, B.; Pooyandeh, M.; Gupta, A.; Motamedi, M.; Marceau, D. Modelling interactions between land use, climate, and hydrology along with stakeholders' negotiation for water resources management. *Sustainability* **2017**, *9*, 2022. [CrossRef]
69. Tangible Landscape. Available online: Tangible-landscape.github.io/ (accessed on 10 May 2020).
70. Souffront Alcantara, M.A.; Kesler, C.; Stealey, M.; Nelson, E.J.; Ames, D.P.; Jones, N.L. Cyberinfrastructure and web Apps for managing and disseminating the National Water Model. *J. Am. Water Resour. Assoc.* **2017**. [CrossRef]
71. Grigg, N.S. Water governance: From ideals to effective strategies. *Water Int.* **2011**, *36*, 799–811. [CrossRef]
72. Blodgett, D.L.; Read, E.K.; Lucido, J.M.; Slawecki, T.; Young, D. An analysis of water data systems to inform the Open Water Data Initiative. *J. Am. Water Resour. Assoc.* **2016**, *52*, 845–858. [CrossRef]
73. Lember, V.; Brandsen, T.; Tõnurist, P. The potential impacts of digital technologies on co-production and co-creation. *Public Manag. Rev.* **2019**, *21*, 1665–1686. [CrossRef]
74. Casiano Flores, C.; Crompvoets, J. Assessing the governance context support for creating a pluvial flood risk map with climate change scenarios: The Flemish subnational case. *ISPRS Int. J. Geo Inf.* **2020**, *9*, 460. [CrossRef]

Article

Quantifying the United Nations' Watercourse Convention Indicators to Inform Equitable Transboundary River Sharing: Application to the Nile River Basin

Yared Gari [1,*], Paul Block [2], Getachew Assefa [3], Muluneh Mekonnen [4] and Seifu A. Tilahun [1]

1 Faculty of Civil and Water Resources Engineering, Bahir Dar Institute of Technology, Bahir Dar University, P.O. Box 26, Bahir Dar, Ethiopia; satadm86@gmail.com
2 Department of Civil & Environmental Engineering, University of Wisconsin-Madison, 1415 Engineering Drive, Madison, WI 53706, USA; paul.block@wisc.edu
3 School of Architecture, Planning and Landscape, University of Calgary, 2500 University Drive NW, Calgary, AB T2N 1N4, Canada; gassefa@ucalgary.ca
4 Department of Resource Management Program, Alberta Environment and Parks, Calgary, AB T2E 7L7, Canada; Muluneh@kth.se
* Correspondence: yaredbacha@gmail.com; Tel.: +251-912-607-103

Received: 10 June 2020; Accepted: 26 August 2020; Published: 8 September 2020

Abstract: East African riparian countries have debated sharing Nile River water for centuries. To define a reasonable allocation of water to each country, the United Nations' Watercourse Convention could be a key legal instrument. However, its applicability has been questioned given its overly generalized guidance and non-quantifiable factors. This study identified and evaluated appropriate indicators that best describe reasonable and equitable principles and factors detailed under Article 6 of the convention in order to allocate Nile River water among the states. Potential indicators (n = 75) were defined based on multiple sources that can address conflicting interests specific to this basin context. A questionnaire based on these indicators was developed and distributed to 215 prominent experts from five professional groups on five continents. To analyze the presence of agreements or disagreements within and outside of the basin, as well as differences across expert groups, a k-mean clustering analysis and statistical tests (ANOVA and t-test) were employed. The results imply agreement on 75% of the proposed indicators by all experts across all continents. However, a significant difference in identifying the importance and relevance of many indicators between experts from Egypt and other countries was evident. This study thus demonstrates how the UN watercourse convention principles can be quantified and applied to transboundary water allocation, and ideally lead to informed discourse between basin countries in conflict.

Keywords: equitable water sharing; UN watercourse convention; international and transboundary rivers; Nile River basin

---

1. Introduction

Conflicts over transboundary river sharing due to increases in demand from growing populations and urbanization to produce food and energy are rapidly growing in many basins [1]. Additionally, the impact of climate change is threatening the supply availability of water resources to a significant degree in many locations [2,3]. This imbalance of demand and supply can lead to water insecurity and drive riparian states to engage in unilateral development rather than follow shared water principles, particularly when no prior agreements exist. Clearly, development plans that benefit a given state and potentially result in reduced water availability to other riparian states may be seen as a threat.

For this reason, disputes arise among basin countries either to secure their water quota or maintain the status quo. The current conflict over utilization of the Nile River between Ethiopia and Egypt is a prime illustration of this situation.

To settle such kind of controversies associated with border crossing rivers water utilization, the necessity of properly executing international law principles is unquestionable [4]. Among many legal instruments discussed in the next section, the recently ratified and enforced "the 1997 UN Convention on the Law of Non-navigational Uses of International Watercourses (UNWC)" can also be applied. This law is divided by seven parts that contain 37 articles. Particularly, when Article 5 of Part II vividly states the principle of equitable and reasonable water utilization, Article 6 of the convention details relevant factors that should be taken into account to allocate the water in the absence of agreement between riparian countries [5].

However, as M. Franck [6] stated, although a list of criteria or factors is essential for implementing the principle of equitable and reasonable utilization, the seven factors categorized under Article 6 of the UNWC below lack measurability. These factors are;

1. Geographic, hydrographic, hydrological, climatic, ecological, and other natural characteristics;
2. Social and economic needs of states;
3. The population dependent on the watercourse in each state;
4. The effects of the use or allocation by one state on other states;
5. Existing and potential uses of the watercourse;
6. Conservation, protection, development, and economic measures of watercourse use and associated costs; and
7. The availability of alternatives, of comparable value, to a particular planned or existing use.

The difficulty of measuring these broad factors in terms of quantity or specific units also casts doubts on its applicability and interpretation [7].

Only a few studies have addressed describing these factors with measurable indicators. For example, Beaumont [8] suggested two indicators, namely relative flow contribution and prior appropriation, to apply the principle of equitable and reasonable water sharing on transboundary rivers. Ziad and Bassam [9] also proposed nine indicators for the Jordan River basin to allocate water between Israel, Jordan, Palestine, Syria, and Lebanon. With the addition of water quality and ecological variables, a study by Kampragou, et al. [10] proposed 13 additional indicators for equitable water allocation in the Nestos River basin. Although the indicators from these studies provide insight into describing factors in Article 6 of the UNWC, other studies have indicated a lack of consensus among multidisciplinary experts on proposed indicators. Findings from a study by Fariba, et al. [11] on the Sirwan-Diyala River between Iran and Iraq also lacked showing the adequacy of the indicators, both in terms of incorporating the conflicting interests of watercourse states and the scope of all factors.

From the above studies, one can see that the indicators suggested by individual studies are prone to subjective interpretation and open to criticism. The consideration of factors such as environmental, social, economic, and political interests of states have also been disregarded in some studies like that of Beaumont [8]. Moreover, the applicability of specific indicators also varies between locations, and applying only a few variables may not be sufficient to fully consider all dynamics.

Therefore, for a river basin such as the Nile, where only two downstream countries fully control the water based on the 1929 and 1959 colonial treaties and disagreements escalate from time to time [12–17], implementing the principle of equitable and reasonable utilization by identifying measurable indicators is warranted. If not, the Declaration of Principles (DoP) agreed upon in 2015 by Egypt, Ethiopia, and Sudan to utilize the river based on equitable and reasonable sharing remains unfunctional [18].

For this purpose, this study identified and evaluated appropriate indicators for factors mentioned in the UNWC to allocate the Nile River in an equitable and reasonable manner among watercourse states.

2. Background of Legal Instruments

Following the industrial revolution, as countries started scrambling to meet their ever-increasing water demand, international institutions were required to develop water laws to govern transboundary rivers [19]. Two disparate doctrines were predominantly cited by riparian countries depending on their position. Upstream countries voiced their "absolute sovereign" right to use a river without any restriction, whereas downstream states claimed their right based on the "absolute territorial integrity" principle. Reconciling these two opposing views was difficult, leading to multiple transboundary water allocation conflicts [20]. In 1956, Dubrovnik developed the "limited sovereignty and territorial integrity" principle, which later became a base for the emergence of the principle of equitable and reasonable utilization. In 1966, the International Law Association (ILA), a scholarly but nonauthoritative institution, adopted this principle into the Helsinki rule. This guiding rule became one of the fundamental legal instruments for managing transboundary rivers. Specifically, Article 4 of the rule outlines 11 factors necessary for implementing the principle of equitable and reasonable utilization of international rivers. Though many treaties and agreements were based on this rule, neither scholars in the ILA nor the document are recognized by all countries, leaving the agreements nonbinding [19]. The same is true for the 2004 "Berlin rule," which was extended after the Helsinki rule by ILA. Although the rule is comprehensive in accounting international environmental law, human rights law and the humanitarian law related to management of all kind of water bodies. Similarly, since countries do not agree to be governed by this rule, it is not effective.

Given the increasing number of transboundary water conflicts, the UN General Assembly assigned legal experts nominated by various countries to carefully evaluate international watercourse laws, which eventually became codified in December 1970. Twenty-seven years later, the "UN Convention on the Law of Non-navigational Uses of International Watercourses (UNWC)" was adopted by the general assembly [21]. Whereas the convention also accepted most of the Helsinki rules, major modifications were made including restricting navigational use, the separation of surface and ground water resources, and condensing the original 11 factors into 7. After the ratification process, the UNWC entered into force operationally in 2014, now serving as a key legal instrument used by the international court of justice (ICJ) to resolve international water disputes [20,22]. The heart of the document is Articles 5–7 detailing the principle of equitable and reasonable utilization.

To highlight the abstract meaning and application of the principle of "equitable and reasonable utilization" mentioned under Article 5 [23], Rieu-Clarke, et al. [5] stated that from both procedural and substantive points of view, the term "equity" in international laws implies fairness and justice. Even when UNWC did not exist in its current position, ICJ was applying this interpretation on many international cases, including river basins such as Danube [24,25]. When the procedural interpretation focuses on the right of sovereign states, the substantive dimension seeks to ensure distributive justice. However, "equity" does not simply imply dividing the resource into equal portions [26]. According to McIntyre [27], "equity" can also be viewed from a natural resource allocation perspective. This includes a matter of ensuring certain levels of fairness between developed and developing countries as well as between current and future generations [28,29]. An example of this is the 1992 Rio Declaration [30]. Among 27 principles, principle-16 of the declaration intended to guide countries future sustainable development based on the "polluter pays principle." It established a modality in which developed countries support the green economy of developing countries to balance the harm caused by excessive emission of pollutant gases. Similarly, the term "reasonable" refers to the rationale behind the process of balancing various conflicting uses by states, including the state's development level and other external forces [5]. Again, reasonability does not necessitate achieving optimum allocation or utilization of advanced technology. Rather, it is a relative consideration of current and future contexts without compromising fairness and justice.

Although this principle is fully integrated into the Nile basin's Cooperative Framework Agreement (CFA) and Declaration of Principles (DoP), the CFA lacks a two-third vote necessary for ratification (rejection by Egypt and Sudan), and the DoP agreement only focuses on Ethiopia, Sudan, and Egypt in relation to GERD filling and operation, excluding the six white Nile parties [31,32]. For this reason, the UN convention based on customary international law and applied by the ICJ was utilized as the legal tool in this study.

3. Methodology and Data

3.1. Study Area

The Nile River is one of the longest rivers in the world, traversing 6853 km. It consists of four major water systems, namely the Blue Nile, Tekeze-Atbara, Baro-Sobat, and the White Nile, before forming the main Nile in Sudan and subsequently flowing north through Egypt into the Mediterranean Sea. The Blue Nile, Baro-Sobat, and Tekeze-Atbara originate in the Ethiopian highlands, contributing 86% of annual flow, leaving the White Nile to contribute 14% from the equatorial African lakes region [33]. The basin covers 3.2 million km^2, including 11 riparian states, namely Burundi, the Democratic Republic of Congo, Egypt, Eritrea, Ethiopia, Kenya, Rwanda, South Sudan, Sudan, Tanzania, and Uganda (Figure 1). The basin is home for over 257 million people, with the majority near or below the poverty line [33–35].

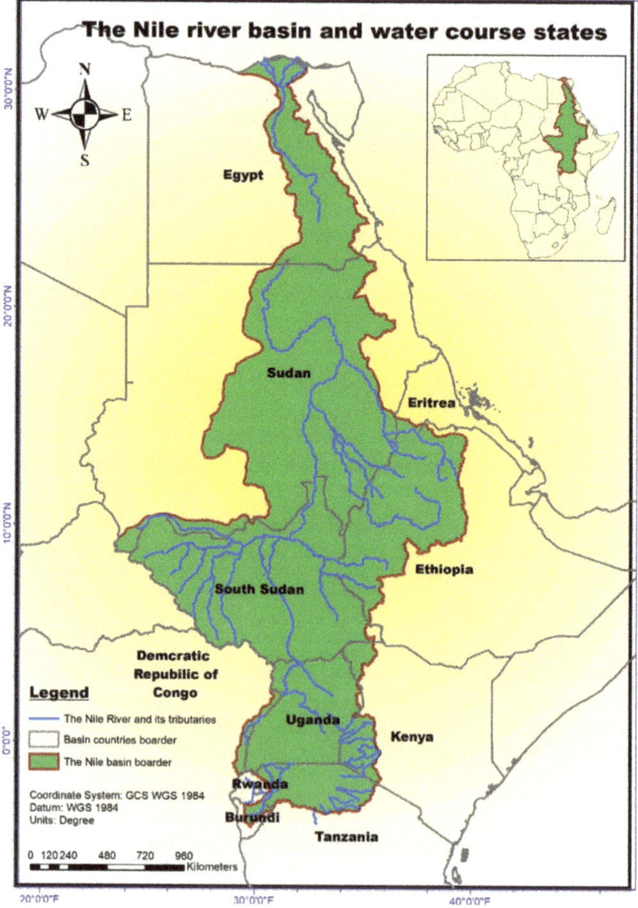

Figure 1. Map of the Nile basin.

3.2. Method

In this study, a cross-sectional analytical research design was applied to assess the level of consensus among experts in different professions and geographic locations on the proposed indicators to define the UN convention's immeasurable factors. For this purpose, an indicator-based questionnaire was selected as the most appropriate and cost-effective method for collecting primary data from relevant experts to capture potential differences in perspectives based on professional background and location (basin vs non-basin, downstream vs upstream, and specific basin country). The selected approach has been applied in similar previous studies. The questionnaire was developed, distributed, and evaluated based on factors from the UNWC, soliciting responses from five continents, to categorize relevance and acceptability of indicators within the Nile region.

3.2.1. Initial Pool of Indicators

Based on Nile River basin hydrographic specifics and socioeconomic characteristics, 75 potential indicators were identified from previous studies and literature, press releases describing negotiations, and websites of international organizations to describe and measure factors contained within Article 6 of the UNWC (Table S1). The main selection criteria considered for these indicators included the availability, affordability, and feasibility to collect and analyze data; measurability; time bounds; the ability to consider watercourse countries claim; the ability to meet obligatory human rights and environmental rules; and the capacity of the indicators to explain unique features of the basin and basin countries. Also, the monitorability and operational usability was considered. The 75 indicators were categorized under the 7 factors previously described. These indicators were given an identification (I) number from 1 to 75 as shown in Table S1. The descriptions of the indicators were also provided in the table, along with information of the potential data source and timeframe for the data collection.

3.2.2. Questionnaire Development and Distribution

The questionnaire was prepared with a Likert scale of 1–5 representing the following for each indicator: 1. Not important, 2. Less important, 3. Neutral, 4. Moderately important, and 5. Very important. It was distributed via email to a total of 215 experts in Africa, Asia, Europe, North America, and South America, with 150 complete responses collected, following a snowball sampling method. The experts represented five professions: Hydrology/water resources, environmental science, law, socioeconomics, and political science. All experts had an educational background at the master's level or above, and worked in universities, ministries, embassies, regional and international agencies and institutes, nongovernmental organizations, and independent consultants (Figures 2 and 3).

To differentiate potential partialities within, outside, and between countries, experts were also grouped as basin and non-basin professionals. To ensure fair representation of experts across all continents, when we planned three responses from each four professions in the nine basin countries due to the difficulty and time it takes to reach out the questionnaire to experts outside of Africa, two responses per five professions was targeted from at least two countries found in the remaining four continents excluding Antarctica and Australia. Accordingly, 135 and 80 experts were targeted from basin and non-basin countries, respectively. The quota of questionnaires among professions also targeted equality. Apart from basin countries, the choice of representative countries from the four continents were made based on their involvement in international water disputes. For example, from Europe, Belgium and the Netherlands were selected given their experience on the Meuse River. Similarly, the USA had negotiations with Canada and Mexico on the Colombia and Rio Grande Rivers, respectively. Additionally, from the Middle East and Asia, Jordan and Israel (Jordan River), Iran and Iraq (Tigris River), as well as China and India (Mekong River), were selected.

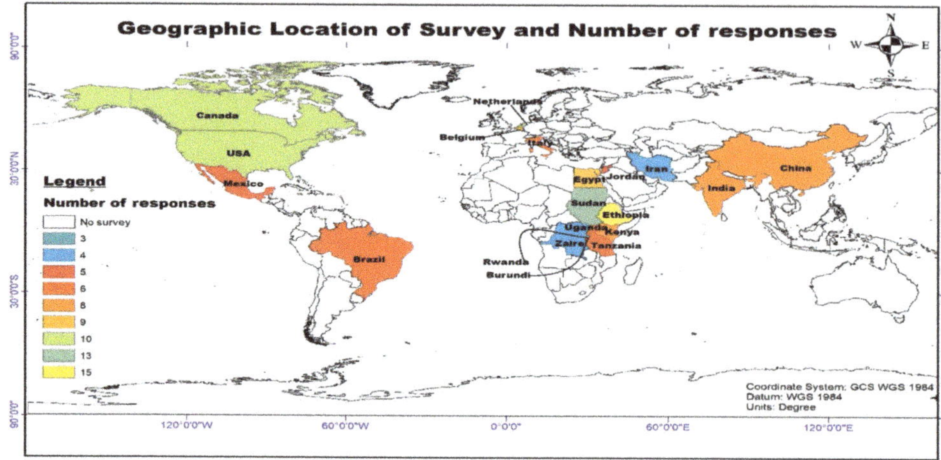

Figure 2. Quantity and spatial distribution of the survey.

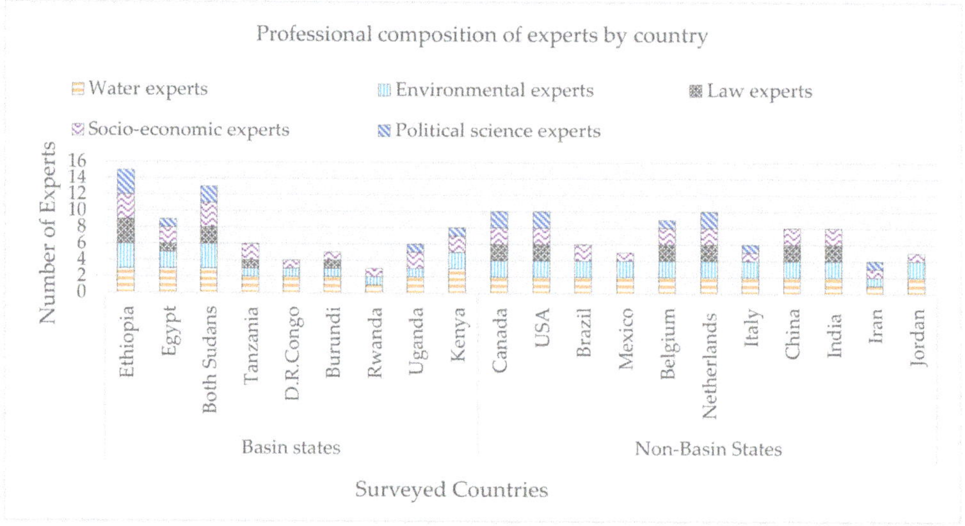

Figure 3. Country, number, and profession composition of experts responded the survey.

3.2.3. Analysis of Survey Data

To avoid overfitting and reduce dimensionality, a principal component analysis (PCA) was employed on all indicators under each factor. A PCA reduces the size of the independent variable set by retaining the maximum variance using fewer dimensions than the original number of indicators. The number of principal components (PCs) retained in this study followed by Formann [36], such that the proportion between the number observations and number variables (v) is greater than 5×2^V. For this study, three PCs were retained under each of the seven factors. All data were confirmed to follow a normal distribution.

After completing the PCA, a k-means clustering algorithm was used to identify groupings of expert responses. K-mean clustering is a partitional clustering approach identifies a user-defined number of clusters (K), which are designated by their means or centroids. To group n-number of observations into a K-number of clusters, this technique uses either the Euclidean or rectilinear distance of these scaled points from the centroids as a measure of similarity. This is performed in an iterative process. First, the numbers of hypothetical clusters (K) based on characteristics as geography, profession and hydrologic position of watercourse states is decided. Second, among the data points, initial centroids are randomly selected. The number of these randomly selected observations (an associated starting centroid) are, by default, equal to the number of assumed clusters. Third, the Euclidean distance between these initial centroids and each data point is calculated using Equation (1). Then, individual observations are classified into k-clusters depending on their minimum distance from the randomly selected centroids. The smaller the Euclidean distance between a given centroid and data points, the higher the probability to be grouped in a similar cluster. However, each time the centroid changes, the cluster of data points also changes. Thus, at each step, centroids are updated by taking the average of the data points that are categorized in the same cluster in the preceding iteration. This continues until a consistent cluster assignment is obtained.

$$d_{u,v} = \sqrt{(u_1 - v_1)^2 + (u_2 - v_2)^2 + \ldots + \left(u_q - v_q\right)^2} \quad (1)$$

where $d_{u,v}$ is Euclidean distance between a given centroid and variable u and v, and $1,2 \ldots q$ are data points or observations of each variables.

Accordingly, we first set the number of hypothesized clusters at $K = 2$ for groupings between basin and non-basin or upstream and downstream, $K = 4$ for grouping among basin states, and $K = 5$ for grouping by profession. Given these specified number of clusters, the algorithm iterated to identify patterns of differences or similarities that existed in the survey data. However, because the ideal number of clusters for a given dataset can be different from what is anticipated, an optimal number of clusters was detected using an elbow method for proofing our assumptions. Repeating the steps above, all variances (within sum of squared errors (SSE)) corresponding to each K values from 2 to $(n - 1)$ were calculated, where n is the number of observations. Then, by plotting (K vs. SSE), a point marking the approximate location where a rapid decline in the slope of the variance ends and began to flatten (forming an elbow shape) was noted. At this point, because the rate of change in the variance was quite small for additional clusters, the corresponding number of clusters (K-value) was selected as appropriate. This also allowed a k-means cluster obtained by an optimal K-value to verify the initially selected number of clusters. Further, the number of respondents categorized in each group from each country was also used as a check to examine the existence of expected grouping in the survey responses.

Although the k-mean clustering identifies patterns and reasons for grouping in the data, it cannot determine the degree of difference between groupings, thus Analysis of Variance (ANOVA) was included to evaluate potentially statistically significant differences among basin countries, and a t-test was used to identify potentially significant differences between, within, and outside of the basin countries for individual indicators. To be statistically significant, results may surpass the 95% confidence level (p-value < 0.05).

3.2.4. Final Indicator Selection

The selection of the final set of indicators was based on both statistical significance and the percentage of responses with a Likert score of 5 (very important) (Figure 4). All indicators were classified into levels of consensus according to the following:

(1) If there is no statistically significant difference among basin countries and between basin and non-basin countries, regardless of the percentage of experts selecting 'very important,' then the degree of consensus for that indicator is considered high;

(2) If there is a statistically significant difference either among basin countries or between basin and non-basin countries, but the percentage of experts selecting 'very important' is >50%, then the degree of consensus is considered moderate;

(3) If there is a statistically significant difference either among basin countries or between basin and non-basin countries, and the percentage of experts selecting 'very important' is <50%, then the degree of consensus is considered low; and
(4) If there is a statistically significant difference among basin countries and between basin and non-basin countries, regardless of the percentage of experts selecting 'very important,' then the degree of consensus for that indicator is considered low.

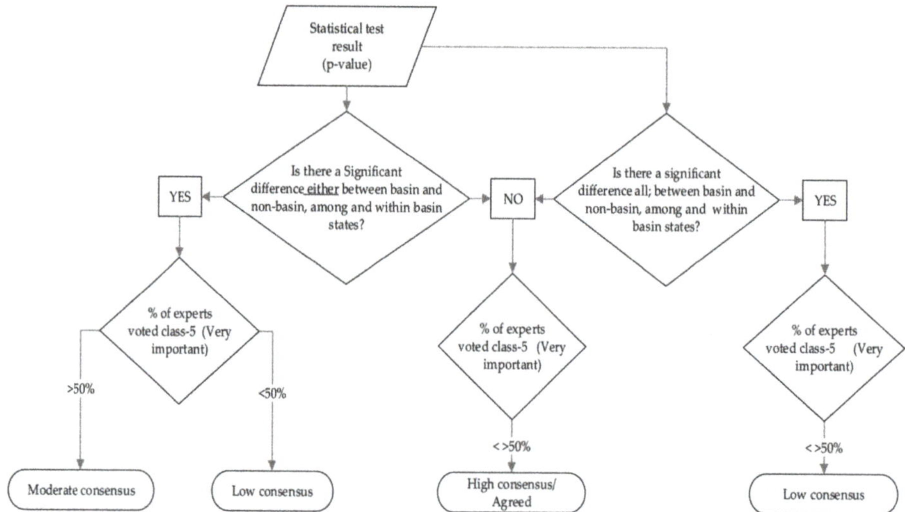

Figure 4. Flowchart for identifying the consensus level for each indicator.

4. Results

In the following section, we present results and outcomes from all respondents, analysis comparing among basin countries and basin versus non-basin groupings, and key indicators identified for the Nile Basin that may be used for equitable and reasonable water allocation among the states.

4.1. Responses of All Experts

According to the summary of data (Figures 5 and 6), the response from basin country experts covered all classes from not important to very important (1–5) for most of the indicators. For non-basin countries, this was not the case. This emphasizes that experts from basin countries appeared more divided on most of the indicators than non-basin experts. Yet, basin experts also expressed a common positive inclination for some indicators, including water-food-energy risk index, population without electricity, the relative significance of hydropower, access to drinking water, access to clean cooking, multidimensional poverty, hunger index, existing irrigation demand, and future domestic water demand. Contrarily, although approximately 80% of non-basin experts considered the majority of indicators to be important, there were some exceptions on which they were divided, including ICT index, life expectancy index, cereal yield, and industry (%GDP).

4.2. Comparison among Basin States

The sum of the variance explained from the first three PCs for each factor ranged from 60% to 95%, expressing the scope of agreement and disagreement between basin experts. The clusters based on these PCs for each factor were quite mixed and did not show a clear distinction (Figures 7, A1 and A2). Overall, the influence of the expert's profession (Figure A1) and home country (Figure A2) appears negligible, whereas a country being grouped by hydrologic position (upstream vs. downstream) did indeed illustrate clearer clusters (Figure 7). This was particularly clear from factors 1 to 5, but less so for factors 6 and 7 (costs of conservation and protection, and the availability of alternative uses (Figure 7f,g), where a stronger similarity was observed). Thus, not all indicators necessarily imply a difference in opinion between upstream and downstream states.

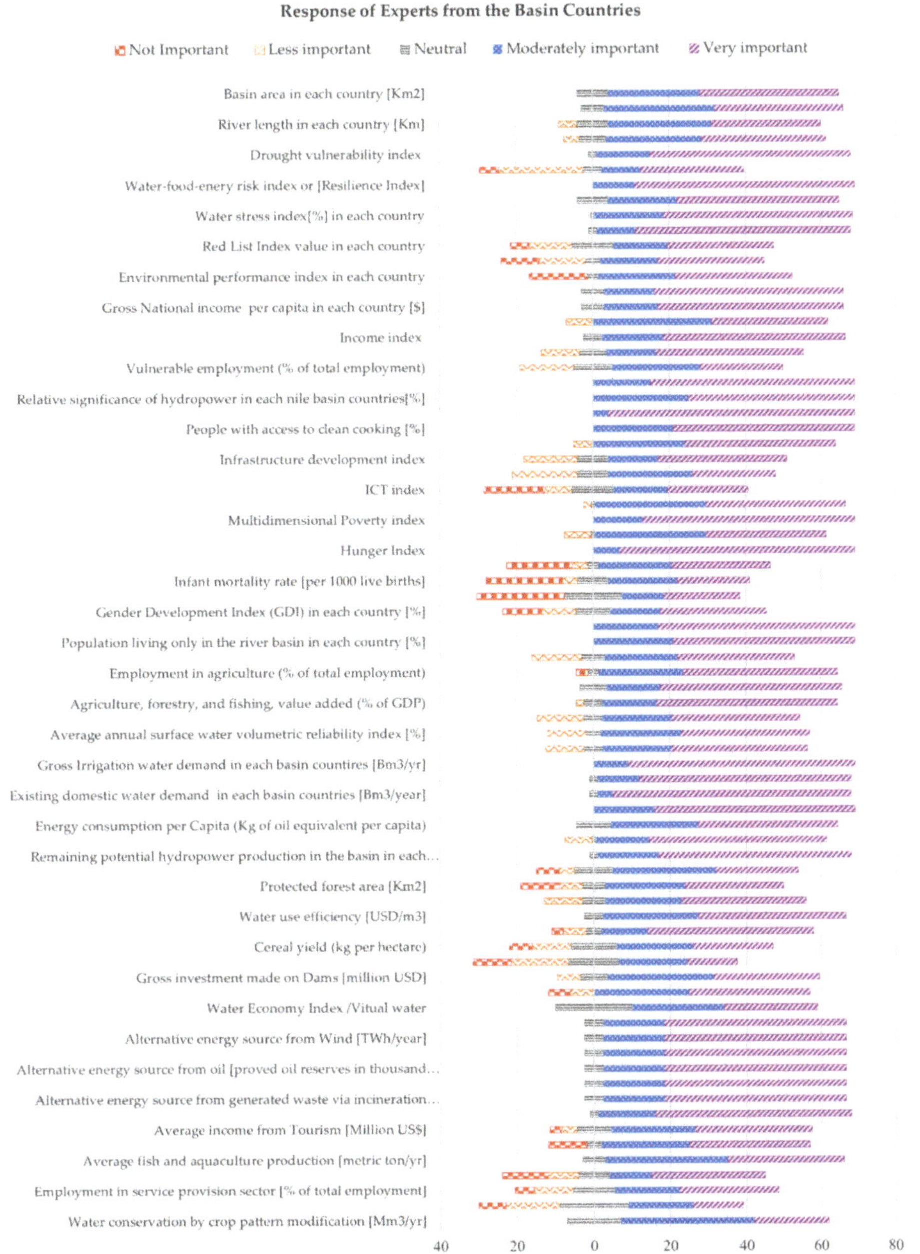

Figure 5. A percentage summary of survey responses for indicators from basin state experts.

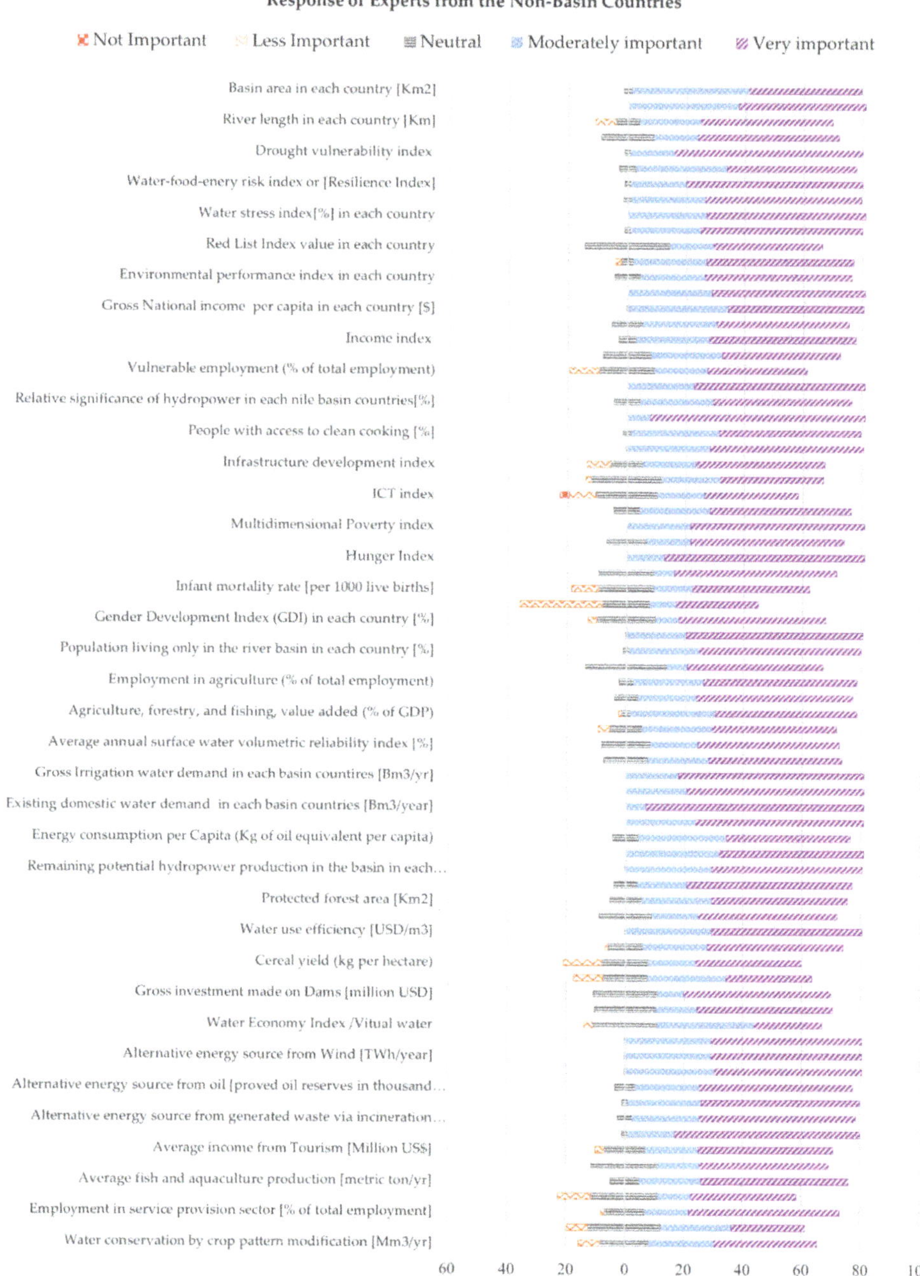

Figure 6. A percentage summary of survey responses for indicators from non-basin state experts.

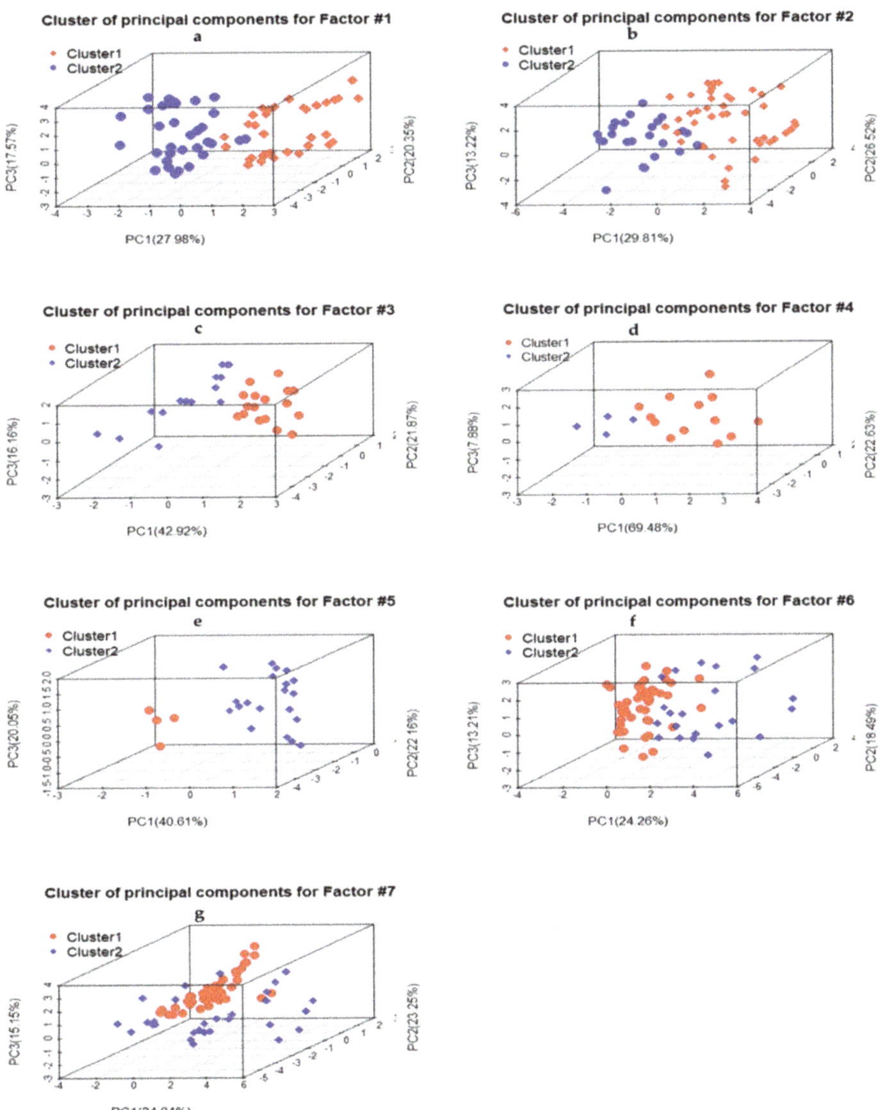

Figure 7. A cluster based on hydrologic position of basin states (i.e., cluster 1 and cluster 2 represent upstream and downstream countries, respectively), where (**a**) is experts response division on indicators under factor-#1—called geography, hydrology, ecology, and natural features, (**b**) factor-#2—socio-economic needs of basin states, (**c**) factor-#3—the population dependent on the watercourse, (**d**) factor-#4—the effects of water use, (**e**) factor-#5—existing and potential uses, (**f**) factor-#6—costs of conservation and protection, and (**g**) factor-#7—availability of alternative and comparable values and uses.

Beyond the confirmation of optimal number of clusters to be ($k = 2$), the number of experts falling into each distinct cluster based on hydrologic position is also insightful (Figure 8). Sudan and Egypt were visibly similar for indicators focused on factor 6 (costs of conservation and protection) and factor 7 (availability of alternative uses). However, for factors 1–5, most Sudanese experts' responses more closely matched upstream expert opinions, leaving a relatively clear separation between Egypt versus other basin states. Although these conclusions can be drawn at the factor scale, broad dissimilarities across all indicators listed under each factor are not necessarily evident. Rather, it is typical only a few indicators under each factor that strongly influence the division by hydrologic position. These specific indicators can be identified through the proposed statistical tests.

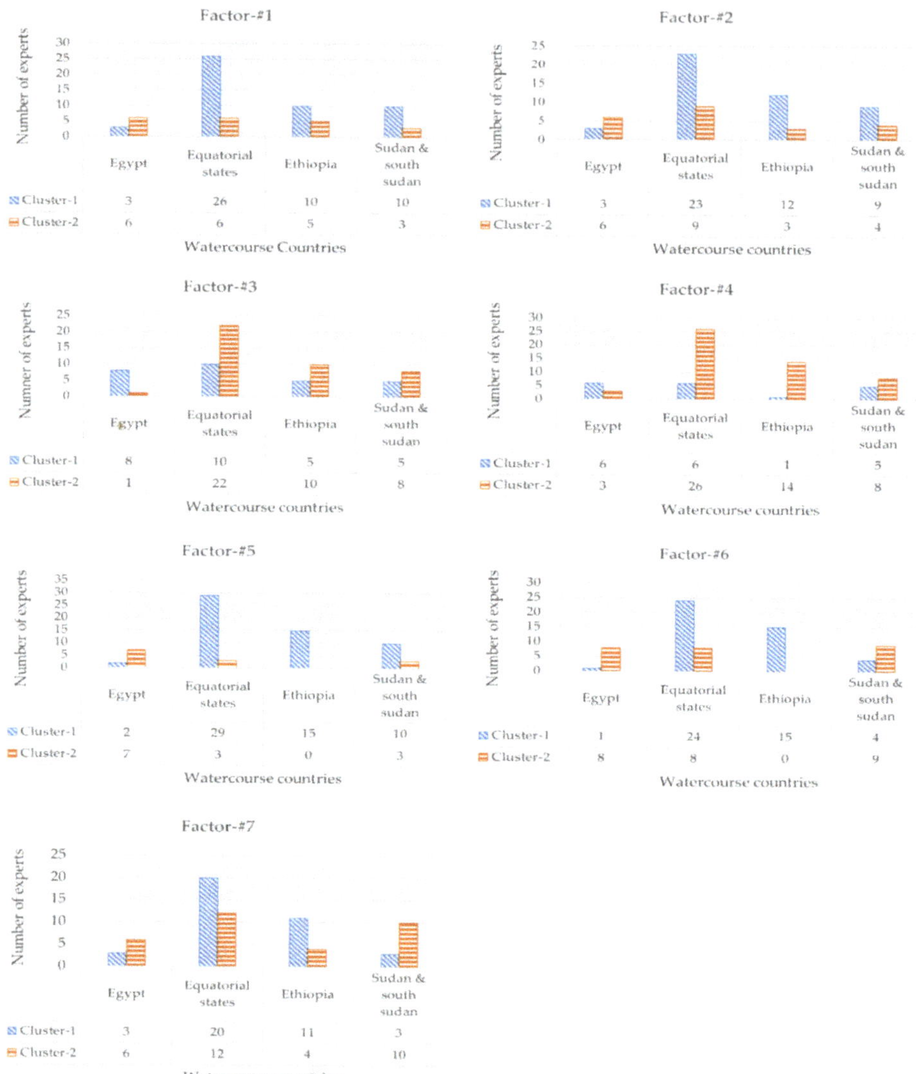

Figure 8. The number of experts from upstream and downstream states in each cluster, organized by factor. Where factor-#1 is geography, hydrology, ecology, and natural features, factor-#2 is socio-economic needs of basin states, factor-#3 is the population dependent on the watercourse, factor-#4 is the effects of water use, factor-#5 is existing and potential uses, factor-#6 is costs of conservation and protection, and factor-#7 is availability of alternative and comparable values and uses.

Whereas a general agreement exists on 67 indicators among basin countries, a significant difference exists for eight indicators among basin states as well as two indicators between Egyptian experts (Table S2). Apart from Egypt's experts, there were no significant differences detected within other basin states. The eight indicators which resulted in a significant difference based on hydrologic position included the average drought-affected people per year in each country (I-6), population living below the income poverty line (I-30), population growth rate (I-38), wetland area (I-56), estimated cost to conserve erosion hot spot areas (I-58), virtual water (I-61), revenue and job opportunity from ports (I-74), and water conservation by crop pattern modification (I-75). The two indicators for which Egyptian experts also significantly disagreed within themselves included the average drought affected people per year in each country (I-6) and estimated cost to conserve erosion hot spot areas (I-58).

4.3. Comparison between Basin and Non-Basin Countries

To compare findings from the within-basin assessment, and particularly to understand the degree to which outcomes strongly favored national interests, the same analysis was repeated to examine the level of agreement between basin and non-basin states. Contrary to the within basin states analysis, clusters between groupings of basin and non-basin states did not indicate a distinct separation (Figure 9). Clusters based on experts' professions between basin and non-basin states was also relatively indistinct at the factor scale.

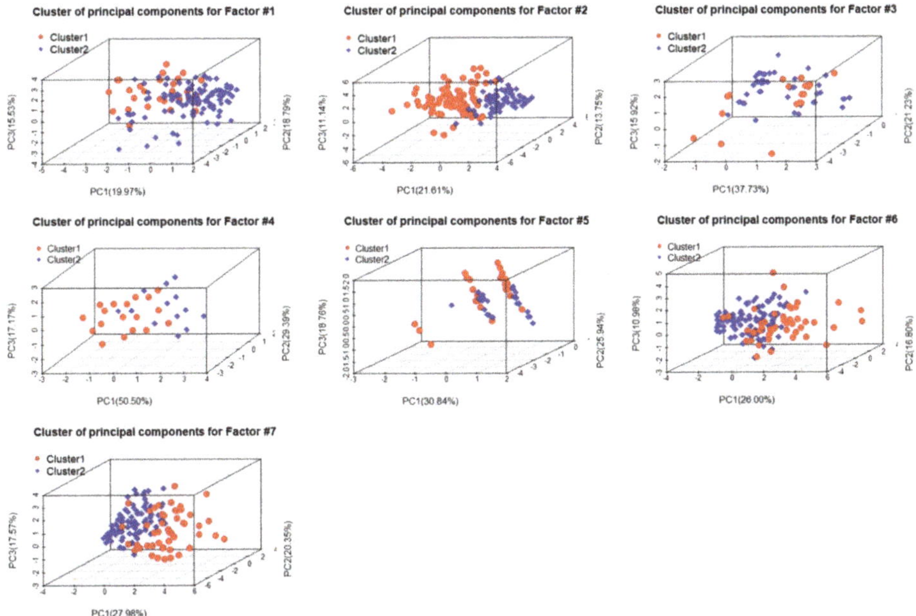

Figure 9. Clusters based on membership in basin (red circles) or non-basin (blue diamonds) states. Where factor—#1 is geography, hydrology, ecology, and natural features, factor-#2 is socio-economic needs of basin states, factor-#3 is the population dependent on the watercourse, factor-#4 is the effects of water use, factor-#5 is existing and potential uses, factor-#6 is costs of conservation and protection, and factor-#7 is availability of alternative and comparable values and uses.

To address possible differences between basin and non-basin states on individual indicators outcomes from a t-test were examined (Table S2). This resulted in basin states experts having a significant difference with non-basin experts on 14 indicators. Three of these indicators were also identified as differences among basin states. These indicators included the average drought-affected people per year (I-6), red list index (I-11), total greenhouse gas emissions (I-12), environmental performance index (I-13), transport index (I-26), ICT index (I-27), education index (I-32), infant mortality rate (I-33), gender development index (I-35), areas exposed for severe soil erosion (I-52), protected forest area (I-53), cost to conserve erosion hot spot areas (I-58), employment in service provision sector (I-73), and revenue and job opportunity from ports (I-74),

To investigate the potential sources of these differences, we compared experts' responses by continent and the Middle East region. However, this did not indicate any potential geographic influence on the result. Similarly, the hydrologic position of non-basin countries did not appear to illustrate any distinct differences.

Comparing individual basin states with aggregated non-basin countries, for the 19 indicators (mean value), one or more basin states exhibited a statistically significant difference with non-basin countries (Figure 10), except for red list index value in each country (I-11). Predominantly, the non-basin experts' opinions aligned with upstream states, especially with Ethiopia. The largest gap for most of the indicators was observed between Egypt and non-basin experts. Sudan experts' opinions aligned closely with Egypt for indicators such as I-58 and I-74. However, for the remaining indicators, Sudan responses aligned more closely with upstream states, particularly with equatorial countries.

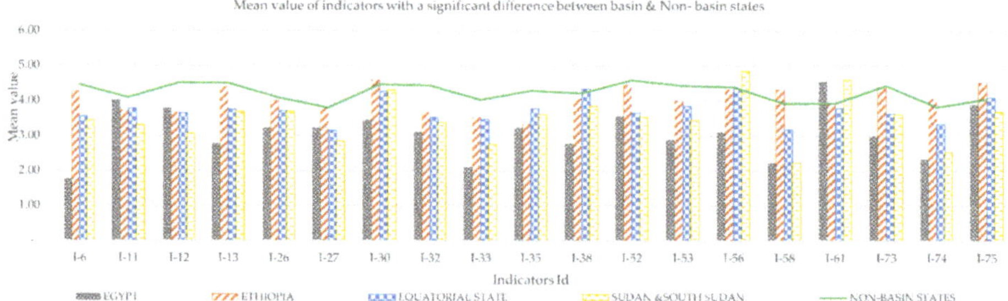

Figure 10. Mean value of all indicators for which individual basin states and non-basin states were statistically significantly different within, among, and between each other. | drought affected people per year (I-6), Red List Index (I-11), greenhouse gas emissions (I-12), environmental performance index (I-13), transport Index (I-26), ICT index (I-27), population living below poverty line (I-30), education index (I-32), infant mortality rate (I-33), gender development index (I-35), population growth rate (I-38), areas exposed for severe soil erosion (I-52), protected forest area (I-53), wetland area (I-56), estimated cost to conserve erosion areas (I-58), virtual water export (I-61), employment in service provision sector (I-73), revenue and job opportunity from ports (I-74), and water saving by crop pattern modification (I-75).

4.4. Important Indicators within Nile Basin

To categorize indicator consensus levels based on expert responses, both the statistical test results and the majority rules detailed in the methods section were applied. Accordingly, for 56 indicators, expert responses exhibited no significant difference, irrespective of their profession and country, and were classified as a high consensus level (Table S2). In contrast, for three indicators, there was no statistically significant agreement either among basin states or between basin and non-basin groups. Thus, they were classified at a level of low consensus. These three indicators included the average drought affected people per year (I-6), estimated cost to conserve erosion hot spot areas (I-58), and revenue and job opportunity from ports (I-74). Nine indicators with a statistically significant disagreement either among basin states or between the basin and non-basin states, yet experts rated the indicators as very important more than50% of the time were classified at a level of moderate consensus. Finally, there were seven indicators for which there was no statistically significant agreement, either among basin states or between basin and non-basin groups. Experts rated the indicators as very important <50% of the time, and they were classified at a level of low consensus. As a result, 10 of the 75 indicators fell into the low consensus category.

5. Discussion

To the best of our knowledge, this study is the first to identify indicators that define factors listed under the 1997 UN Watercourse Convention by involving multidisciplinary experts from five areas of expertise around the globe. Overall, though the majority of 215 experts tended to rate most of the indictors as important, there were also notable differences between indicators. The investigations into possible sources of these differences by clustering analysis indicate neither professional background nor experts' geographic home in terms of basin and non-basin states play a significant role. A clear grouping in data was observed, however, considering experts in upstream versus downstream countries. Still, this does not necessarily imply there are no difference between basin and non-basin or individual countries. Rather, the statistical test outcomes confirmed the existence of a significant difference for 8 (among basin countries) and 11 (between basin and non-basin countries) indicators. In addition, Egyptian experts had a significant difference on 2 of these 19 indicators among themselves.

The mean comparison of these 19 indicators suggests that Egypt experts' preferences are different from other countries. This also seems likely due to the influence of national interests than other scientific justification for water allocation. Considering the socioeconomic conditions of Egypt versus other basin states—with the largest per capita water storage, fewest impoverished people, fixed crop types, and largest port-based income [33,37–39]—the influence of nationalistic bias is not unexpected. Sudan exhibits a mix of characteristics. Given the high crop water requirements in both Egypt and Sudan as compared to other basin states [40], both are interested in virtual water trade (I-61). Similarly, experts from Sudan and Egypt valued the importance of considering costs to conserve severe erosion (I-58) and opportunities from ports (I-74) less. However, the two countries were distinctly different on the remaining indicators, with Sudan generally aligning more with other basin states (Figure A3–Appendix A). Comparing countries, these indicator preferences were generally aligned with the positions of basin states regarding the Cooperative Framework Agreement, the most recent legal instrument based on the principle of equitable and reasonable utilization still waiting for the required number of ratifications to go into force [31]. Nevertheless, none of the experts from the non-basin countries were observed to side with a particular basin country (Figure A4–Appendix A).

.Therefore, in addition to 24 unique indicators previously applied in different studies to inform fair share of basin states [8–11], this study introduced 51 additional indicators. Of these, 56 out of the 75 indicators were categorized as highly important, and of the remaining 19 indicators for which significant differences were observed, the level of consensus was labelled as moderate and low (9 and 10), respectively. Out of to the 24 indicators used in previous studies on different river basins, 16 of them were found highly relevant to the Nile basin.

6. Conclusions

The principle of equitable and reasonable utilization agreed to by Egypt, Sudan, and Ethiopia in their 2015 Declaration of Principles (DoP) does not readily allow for quantification of water sharing due to limitations of international laws in detailing measurable criteria for factors listed in this document and the UNWC guidelines. This study identified and evaluated basin-specific indicators by engaging professionals from basin and non-basin states. As observed from the summary of survey result, although experts from different geographic locations have divided responses on the importance of 75 proposed indicators, about 60% and 80% of basin and non-basin professionals, respectively, had a tendency to consider the majority of indicators to be relevant. The statistical tests revealed that experts grouped as basin versus non-basin and downstream versus upstream states had a significant difference among, between, and within countries on 19 indicators. The findings also illustrate that the major differences in assigning importance levels to proposed indicators mainly occurred between Egypt and other countries, even more so than between basin and non-basin states. Furthermore, a clustering analysis indicates that these differences were likely more a result of national interests rather than profession background.

Though the position of Sudan varies from indicator to indicator, the findings of this study are generally similar to the position that basin countries have on the Cooperative Framework Agreement and current disagreements over the Grand Ethiopian Renaissance Dam. In conclusion, out of 75 indicators, multidisciplinary experts identified 56 as highly relevant indicators, while only 9 (10) indicators were categorized as moderate (less) relevant. On this basis, given the number of highly important indicators, the degree of disagreement is not interpreted to be as wide as reported in press releases following talks and negotiations discussing political engagements over the use of the Nile River. Rather, the indicators for which there is a high level of consensus can potentially lead to a solid foundation for putting the UNWC into action and pave the way for utilization of the Nile River based on the equitable and reasonable principle. Moreover, in addition to facilitating the allocation of water between riparian countries, breaking down these broad factors into measurable indicators could help legal and water resource policy makers to resolve disputes and mitigate controversial issues in the river basin. However, as highlighted in the results, since competing interests of the basin countries potentially influence expert's judgment and these interests vary from basin to basin, all indicators evaluated in this study for the Nile River basin may not be appropriate for other basins, as local context must be considered. Even within the same basin, regularly gauging expert opinions about indicators may also be important as countries and conditions change.

As the number of experts surveyed increases from all locations, the categories of moderately important indicators may change. However, we deem the 56 highly important indicators as adequate to improve measurability of the UNWC factors, capture the unique features of the Nile basin, and accommodate conflicting interests of states. Beyond laying a foundation for operationality of the UNWC, the outcomes of this study pave the way for the quantification of water allocations to riparian countries. Future work should address investigating the priority or weight of individual indicators for use in water resources modelling.

Supplementary Materials: The following are available online at http://www.mdpi.com/2073-4441/12/9/2499/s1, Table S1: Factors, Potential indicators and their description, Table S2: Cumulative percentage vote of indicators and the level of consensus among experts.

Author Contributions: Conceptualization, Y.G. and S.A.T.; Data curation, Y.G.; Formal analysis, Y.G.; Funding acquisition, G.A. and S.A.T.; Investigation, Y.G.; Methodology, Y.G. and S.A.T.; Project administration, S.A.T.; Resources, P.B., G.A. and S.A.T.; Software, Y.G.; Supervision, P.B. and S.A.T.; Validation, P.B. and S.A.T.; Visualization, Y.G.; Writing—original draft, Y.G.; Writing—review & editing, P.B., G.A., M.M. and S.A.T. All authors have read and agreed to the published version of the manuscript.

Funding: This work was financially supported by a collaborative partnership between the University of Calgary, Canada and Bahir Dar University (BDU), Ethiopia.

Acknowledgments: The authors would like to thank University of Wisconsin-Madison for facilitating data collection and research visit.

Conflicts of Interest: The authors declare no conflict of interest.

Appendix A

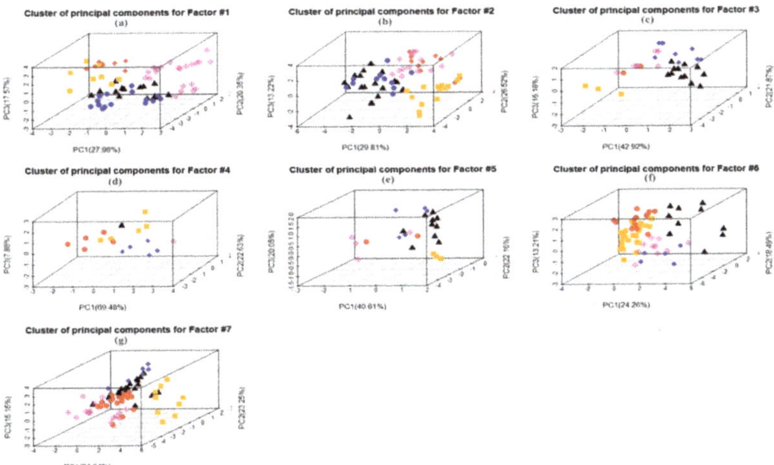

Figure A1. A cluster based on basin experts' professions, in which black triangles are hydrologists, yellow rectangles are socioeconomics, purple crossed circles are environmentalists, red circles are political scientists, and blue diamonds are law experts, where (**a**) are experts response divisions on indicators under factor-#1—natural features, (**b**) factor-#2—socioeconomic needs of basin states, (**c**) factor-#3—the population dependent on the watercourse, (**d**) factor-#4—the effects of water use, (**e**)factor-#5—existing and potential uses, (**f**) factor-#6—costs of conservation & protection, and (**g**) factor-#7—availability of alternative and comparable values.

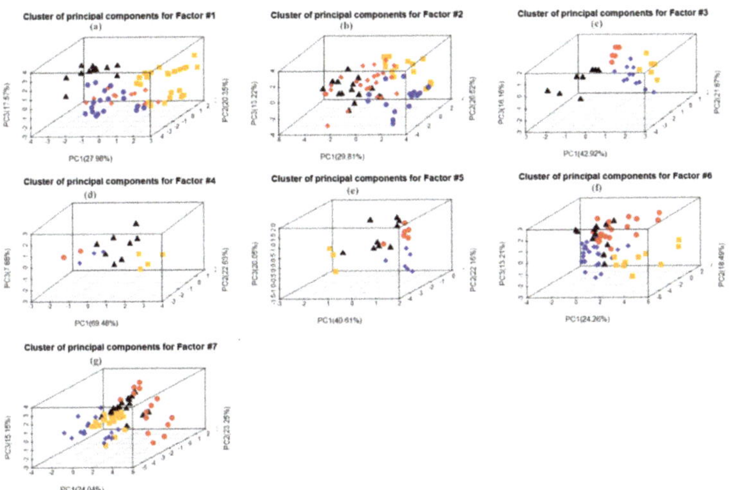

Figure A2. A cluster based on individual basin experts' country of origin, in which black triangles are Egyptian, yellow rectangles are Ethiopians, red circles are equatorial states, and blue diamonds are Sudanese, where (**a**) is experts response division on indicator under factor-#1 called natural features, (**b**) factor-#2—socioeconomic needs of basin states, (**c**) factor-#3—the population dependent on the watercourse, (**d**) factor-#4—the effects of water use, (**e**)factor-#5—existing and potential uses, (**f**) factor-#6—costs of conservation and protection, and (**g**) factor-#7—availability of alternative and comparable values.

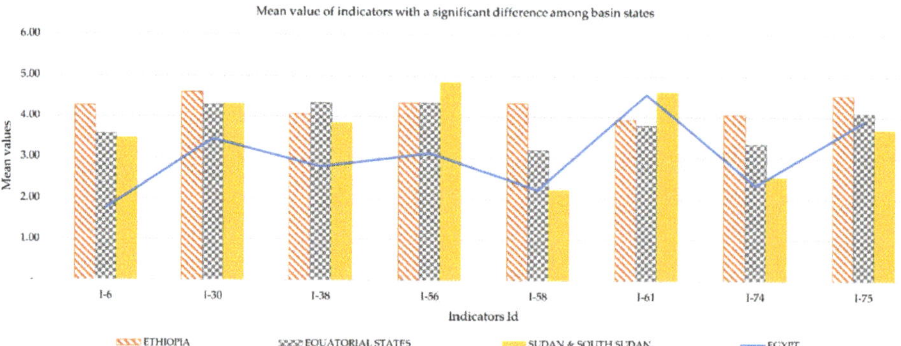

Figure A3. Mean value of indicators for which basin states were statistically significantly different.

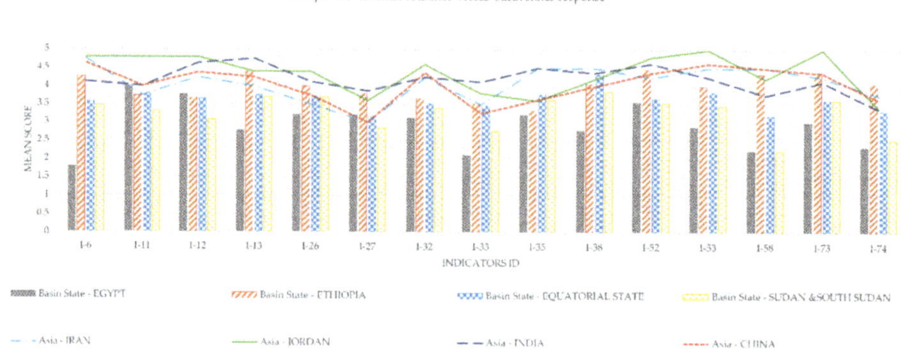

Figure A4. The response of Arab league and middle east countries as compared to basin states experts.

References

1. Boretti, A.; Rosa, L. Reassessing the projections of the World Water Development Report. *NPJ Clean Water* **2019**, *2*, 1–4. [CrossRef]
2. Roth, V.; Lemann, T.; Zeleke, G.; Subhatu, A.T.; Nigussie, T.K.; Hurni, H. Effects of climate change on water resources in the upper Blue Nile Basin of Ethiopia. *Heliyon* **2018**, *4*, e00771. [CrossRef] [PubMed]
3. Abbaspour, K.C.; Faramarzi, M.; Ghasemi, S.S.; Yang, H. Assessing the impact of climate change on water resources in Iran. *Water Resour. Res.* **2009**, *45*, 1–16. [CrossRef]
4. Yihdego, Z.; Rieu-Clarke, A. An exploration of fairness in international law through the Blue Nile and GERD. *Water Int.* **2016**, *41*, 528–549. [CrossRef]
5. Rieu-Clarke, A.; Moynihan, R.; Magsig, B.-O. *Un Watercourses Convention User's Guide*; IHP-HELP Centre for Water Law: Dundee, UK, 2012.
6. Salman, S.M.A. The Helsinki Rules, the UN Watercourses Convention and the Berlin Rules: Perspectives on International Water Law. *Int. J. Water Resour. Dev.* **2007**, *23*, 625–640. [CrossRef]
7. Muhammad, M.R. Principles of international water law: Creating effective transboundary water resources management. *Int. J. Sustain. Soc.* **2009**, *1*, 207–223.
8. Gabriel, E. Development of International law and the united nation watercourse convention. In *Development of International Law and the UN Watercourse Convention, in Hydropolitcs in the Developing world: A Southern African Perspective*; University of Pretoria: Pretoria, South Africa, 2002; pp. 89–96.
9. Tekuya, M. The Egyptian Hydro Hegemony In The Nile Basin: The Quest For Changing The Status Quo. *J. Water Law* **2018**, *26*, 10–20.

10. Jonathan, L.; Mark, G. Equity in Transboundary Water Law: Valuable Paradigm or Merely Semantics. *Colo. J. Int. Environ. Law Policy* **2006**, *17*, 89–122.
11. Wouters, P.K. An Assessment of Recent Developments in International Watercourse Law through the Prism of the Substantive Rules Governing Use Allocation (in English). *Nat. Resour. J.* **1996**, *36*, 417–440.
12. Anikó Raisz, J.E.S. Cross border issues of the Hungarian water resources. *Q. J. Environ. Law* **2017**, *7*, 73–98.
13. Dellapenna, J.W. The customary international law of transboundary fresh waters. *Int. J. Glob. Environ.* **2001**, *1*, 264–305. [CrossRef]
14. McIntyre, O. Utilization of shared international freshwater resources–The meaning and role of "equity" in international water law. *Water Int.* **2013**, *38*, 112–129. [CrossRef]
15. Rajamani, L. *Differential Treatment in International Environmental Law*; Oxford University Press: New York, NY, USA, 2012.
16. FitzMaurice, M. Intergenerational Trusts and Environmental Protection. By Catherine Redgwell. [Melland Schill Studies in International Law, Juris Publishing, Manchester University Press. 1999. xiv + 200 pp.]. *Int. Comp. Law Q.* **2000**, *49*, 244–245. [CrossRef]
17. Jeffrey, D.K. A Short Guide to the Rio Declaration. *Colo. J. Int. Environ. Law Policy* **1993**, *4*, 119–140.
18. Franck, T.M. *Fairness in International Law and Institutions*; Oxford University Press: London UK, 1995.
19. Bruce, L. Does Article 6 (Factors Relevant to Equitable and Reasonable Utilization) in the UN Watercourses Convention misdirect riparian countries? *Water Int.* **2013**, *38*, 130–145.
20. Beaumont, P. The 1997 UN Convention on the Law of Non-navigational Uses of International Watercourses: Its Strengths and Weaknesses from a Water Management Perspective and the Need for New Workable Guidelines. *Int. J. Water Resour. Dev.* **2000**, *16*, 475–495. [CrossRef]
21. Ziad, A.M.; Bassam, I.S. A Decision Tool for Allocating the Waters of the Jordan River Basin between all Riparian Parties. *Water Resour. Manag.* **2003**, *17*, 447–461.
22. Kampragou, E.; Eleftheriadou, E.; Mylopoulos, Y. Implementing Equitable Water Allocation in Transboundary Catchments: The Case of River Nestos/Mesta. *Water Resour. Manag.* **2006**, *21*, 909–918. [CrossRef]
23. Fariba, A.; Jalal, A.; Ali, M. Modelling Equitable and Reasonable Water Sharing in Transboundary Rivers: The Case of Sirwan-Diyala River. *Water Resour. Manag.* **2017**, *31*, 1191–1207.
24. Collins, R.O. *The Waters of the Nile: Hydropolitics and the Jonglei Canal 1900–1988*; Clarendon Press: Oxford, UK, 1990.
25. Abdalla, I.H. The 1959 Nile Waters Agreement in Sudanese-Egyptian relations. *Middle East. Stud.* **1971**, *7*, 329–341. [CrossRef]
26. Lumumba, P.L.O. The interpretation of the 1929 treaty and its Legal Relevance and Implications for the Stability of the Region. *Afr. Sociol. Rev. Rev. Afr. De Sociol.* **2010**, *11*, 10–24. [CrossRef]
27. Lawson, F.H. Egypt versus Ethiopia: The Conflict over the Nile Metastasizes. *Int. Spect.* **2017**, *52*, 129–144. [CrossRef]
28. Obengo, J.O. Hydropolitics of the Nile: The case of Ethiopia and Egypt. *Afr. Secur. Rev.* **2016**, *25*, 95–103. [CrossRef]
29. Mohamed, N.N. Continuous Dispute Between Egypt and Ethiopia Concerning Nile Water and Mega Dams. In *The Handbook of Environmental Chemistry*; Negm, A., Abdel-Fattah, S., Eds.; Grand Ethiopian Renaissance Dam Versus Aswan High Dam; Springer: Berlin/Heidelberg, Germany, 2018.
30. Bayeh, E. Agreement on Declaration of Principles on the Grand Ethiopian Renaissance Dam Project: A Reaffirmation of the 1929 and 1959 Agreements? *Arts Soc. Sci. J.* **2016**, *7*, 1–13.
31. Salman, S.M.A. The Nile Basin Cooperative Framework Agreement: A peacefully unfolding African spring? *Water Int.* **2013**, *38*, 17–29. [CrossRef]
32. Salman, S.M.A. The Grand Ethiopian Renaissance Dam: The road to the declaration of principles and the Khartoum document. *Water Int.* **2016**, *41*, 512–527. [CrossRef]
33. NBI. The Nile Basin Water Resource Atlas. 2016. Available online: http://atlas.nilebasin.org/treatise/nile-basin-water-resources-atlas/ (accessed on 10 May 2019).
34. Abtew, W.; Dessu, S.B. The Nile River and Transboundary Water Rights. In *The Grand Ethiopian Renaissance Dam on the Blue Nile*; Springer International Publishing: Cham, Switzerland, 2019; pp. 13–27.
35. John, O.; Hellen, O.-N.; Lydia, N.; Nancy, L.; Nicholas, W. Exploration of food security situation in the Nile basin region. *Dev. Agric. Econ.* **2011**, *3*, 74–285.
36. Formann, A.K. *Die Latent-Class-Analyse: Einführung in Theorie und Anwendung*; Beltz: Weinheim, Germany, 1984.

37. Ahmed, A.A.; Ismail, U.H.A.E. *Sediment in the Nile River System*; UNESCO International Hydrological Programme International Sediment Initiative: Khartoum, Sudan, 2008.
38. Grebmer, K.v.; Bernstei, J.; Patterson, F.; Wiemers, M.; Chéilleachair, R.N.; Foley, C.; Gitter, S.; Ekstrom, K.; Fritschel, H. Global Hunger Index The Challenge Of Hunger And Climate Change. Global Hunger Index, Dublin/Bonn2019. Available online: https://www.globalhungerindex.org/results.html (accessed on 10 May 2019).
39. Commercial Ports. Maritime Transport Sector. Available online: http://mts.gov.eg/en/sections/10/1-10-Commercial-Ports (accessed on 12 May 2019).
40. El-Marsafawy, S.M.; Swelam, A.; Ghanem, A. Evolution of Crop Water Productivity in the Nile Delta over Three Decades (1985–2015). *Water* **2018**, *10*, 1168–1170. [CrossRef]

Communication

An Integrative Framework for Stakeholder Engagement Using the Basin Futures Platform

Jackie O'Sullivan [1,*], Carmel Pollino [1], Peter Taylor [2], Ashmita Sengupta [1] and Amit Parashar [1]

[1] CSIRO Land and Water, GPO Box 1700, Canberra, ACT 2601, Australia; carmel.pollino@csiro.au (C.P.); ashmita.sengupta@csiro.au (A.S.); amit.parashar@csiro.au (A.P.)

[2] CSIRO Data61, Sandy Bay, TAS 7005, Australia; peter.taylor@data61.csiro.au

* Correspondence: Jackie.osullivan@csiro.au

Received: 31 May 2020; Accepted: 21 August 2020; Published: 26 August 2020

Abstract: Water resources are under growing pressures globally, and better basin planning is crucial to alleviate current and future water scarcity issues. Communicating the complex interconnections and needs of natural and human systems is a significant research challenge. With advances in cyberinfrastructure allowing for new innovative approaches to basin planning, this same technology can also facilitate better stakeholder engagement. The potential benefits of using digital basin planning platforms for stakeholder engagement are immense; yet, there is limited guidance on how to best use these platforms for more effective stakeholder engagement in water-related issues and projects. We detail our digital platform, Basin Futures, and highlight the potential uses for stakeholder engagement through an integrative framework across different assessment levels. Basin Futures is a web application that is an entry-level modelling tool that aims to support rapid and exploratory basin planning globally. As a cloud-based tool, it brings together high-performance computing and large-scale global datasets to make data analysis accessible and efficient. We explore the potential use of the tool through three case studies exploring agricultural development, transboundary water-sharing agreements and allocating water for environmental flows.

Keywords: river basin planning; digital platforms; stakeholder engagement; integrated water resource management

1. Introduction

Water resources are under growing pressures globally, with nearly 80% of the world's population are exposed to a high level of threats stemming from water insecurity [1–3]. Water insecurity is manifested through physical shortages, failure of institutions or lack of infrastructure [4]. Shortages in water impact on people's health, livelihoods, ecosystems and the ability to produce food [4–6]. It also impacts a nation's ability to achieve Sustainable Development Goals. Water resource issues are driven by inter-dependencies between hydrological, social, economic and ecological needs in river, lake and aquifer basins. Communicating the complex interconnections and needs of natural and human systems is a significant research challenge [7]. Better water planning is crucial, and if water management habits do not change, the global demand for water could increase by 50% by 2030 [8]. In the face of climate change, population growth, economic development and increased water demand, these inter-dependencies require more integrated approaches to developing and managing water and land resources [9].

The use of technology has advanced significantly in the last 20 years. The expansion in the capabilities of computing power and cyberinfrastructure provides a new approach to address water resource issues and engagement with society [10]. Complex interconnections and unprecedented changes between natural and human systems are recognised as a significant research challenge [7]. There is a growing need for long-term strategic basin planning that crosses sectors and jurisdictions to encapsulate changes in water resources. Innovative and technologically advanced solutions are required to manage water resources and facilitate stakeholder engagement in an accountable and non-discriminatory manner [7,10].

There are numerous economic, environmental and social benefits to be gained from effectively engaging with stakeholders at the basin level in water policies and projects [11–14]. Stakeholder engagement is often undertaken in an ad-hoc and ineffective manner. Implementing a digital stakeholder engagement strategy can dramatically improve stakeholder participation and outcomes [15–17]. The use of digital platforms assists in providing a more transparent and authentic stakeholder engagement program and therefore improving the likelihood of engagement. Digital tools provide accountable and transparent information to reassure the public of the evidence base underpinning water management plans and proposals [18]. This accountability and ability to reach more participants in a non-discriminatory manner aligns with the principles of good water governance of Integrated Water Resource Management (IWRM) [3]. New cloud technologies for the data-intensive world can provide the ability to analyse and integrate the vast and complex historic and current environmental information to manageable levels [19]. Cloud technology can help to visualise issues and options to make it easier to gain creative insights and build collaborations with stakeholders [19].

Alleviating water security challenges requires basic information on water resources [20]. This information includes knowing how much water is available, where it is distributed and how it will change under scenarios of development and climate change. Data and models are often used to address these questions. However, data are fragmented, difficult to access and process and models require significant expense, time to develop, and advanced capability and capacity to use. As a result, it can be difficult and expensive to support the basic information needs to overcome water scarcity and for stakeholders to speak a common language. Basin Futures was developed in response to our experiences working in Australia and internationally on potential water resource developments and complex water management issues. During this time, we noticed users spent a lot of time in data discovery, transformation and understanding complex modelling software. This resulted in limited uptake and the use of tools which limited the level of engagement and participation of stakeholders in water-related issues. We detail our digital observatory platform, Basin Futures, and highlight the potential uses for stakeholder engagement through an integrative framework. We explore the potential use of the tool through three case studies exploring agricultural development, transboundary water-sharing agreements and allocating water for environmental flows.

2. The Basin Futures Platform

The Basin Futures platform was designed based on the needs expressed by catchment managers, policymakers, environmental groups, and scientists. It is designed to lower the barrier to entry for conducting initial assessments of basin water resources and explore future development scenarios. Basin Futures leverages existing data (global and local) to empower decision-makers to understand their opportunities and constraints in managing water resources [21]. The platform integrates global data with models, all packaged in a web application with readily available cloud processing. This reduces costs on multiple fronts: data integration, model setup and integration, and infrastructure. This setup allows quick range-finding scenarios to be run, providing an explorative tool to aid water resource planning discussion. Basin Futures currently has two main modelling components: rainfall-runoff via GR4J [22] to generate the runoff in a basin, and a custom-built reach model for undertaking water management and transfer activities such as water storage, hydropower generation, routing of flows through the basin, demands/consumption and crop modelling (Figure 1). Basin water balance is produced on a daily basis but lumped monthly, with outputs able to be reported on a monthly or annual scale. Future scenarios are able to be run to explore climate change and changing population scenarios. Basin Futures uses climate scenarios based on The Inter-Sectoral Impact Model Inter-comparison Project (ISIMIP) outputs [23]. Enabling users to plan for climate resilient basins by assessing the potential changes in the quantity and timing of runoff, precipitation and streamflow based on multiple future climate scenarios.

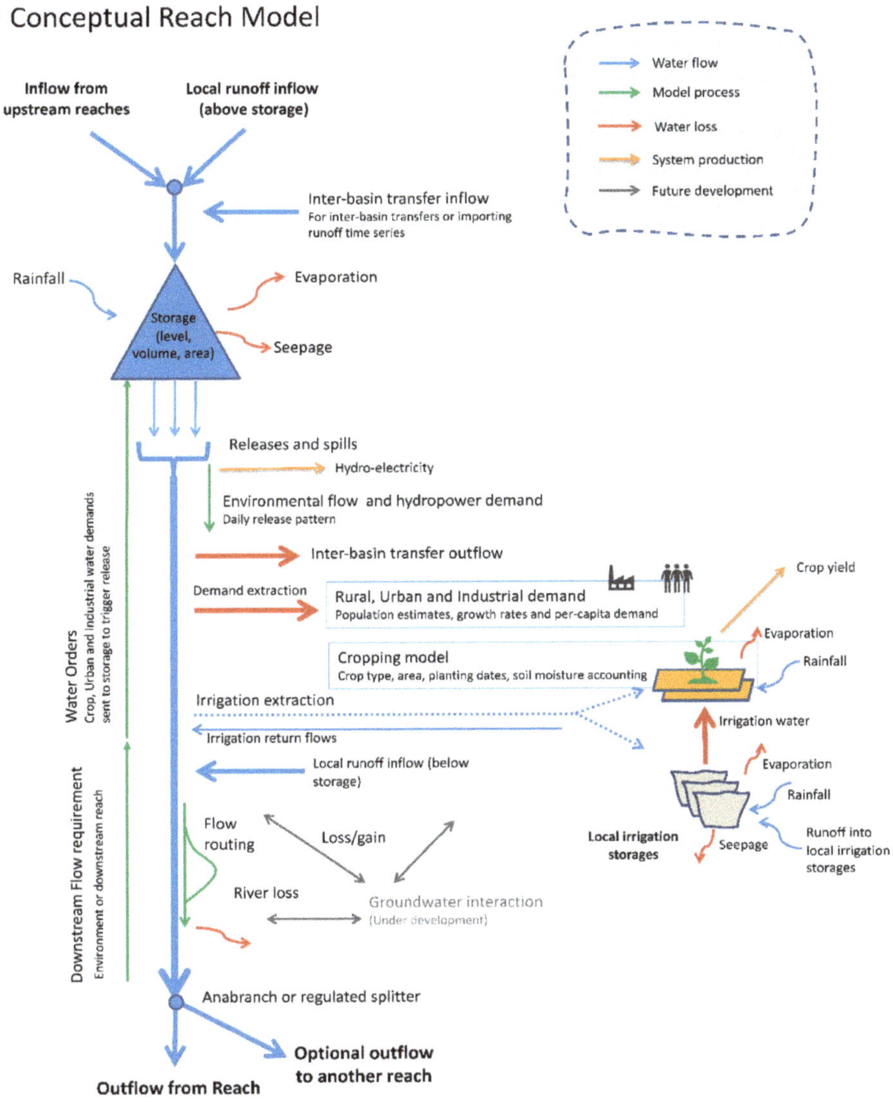

Figure 1. Conceptualisation of the reach model engine underlying Basin Futures.

Our approach focusses on three aspects: global application, easy development of scenarios, and a lower barrier to entry. These key aspects are where Basin Futures differs from existing systems, such as those focused on specific water management areas such as flood forecasting [24,25]. Our approach is similar to a recent web implementation of the Soil and Water Assessment Tool (SWAT) Online [26]; however, we provide scenario development and result assessment. Basin Futures supports users with a robust yet simple modelling framework and can provide a pathway to more sophisticated products through export capabilities. The model workflow provides a consistent and repeatable process of basin water assessments for any basin across the globe (Figure 2). Basin Futures provides an enabling environment for planning, cooperation and participation in water management.

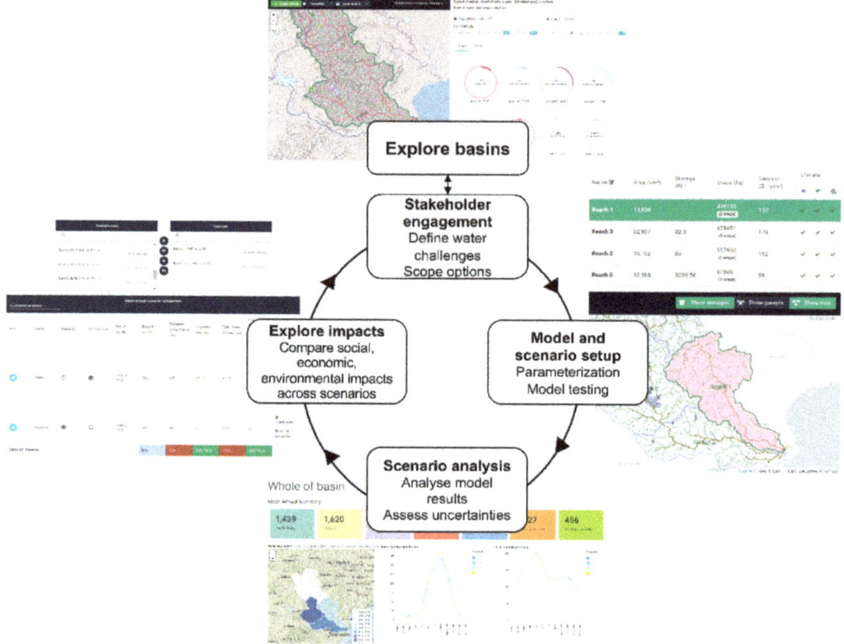

Figure 2. User workflow within Basin Futures.

3. Stakeholder Engagement: Encouraging and Scaling Participation in the Basin Planning Process

Water-related developments or changes require collaboration between a wide range of stakeholder groups from landholders to scientists to policymakers. Depending on the complexity, scale and impacts of the project, the level of stakeholder engagement can range from merely providing information to active partnerships and collaboration. For stakeholders to fully engage, an enabling environment for effective, fit-for-purpose and outcome-oriented stakeholder engagement should be provided [27]. Digital platforms can be used as a common platform between stakeholders to contribute data, knowledge and understanding and visualisation of an issue [7,10].

Digital platforms can be used to reach a wider stakeholder audience to encourage participation in water-related planning and development. Digital platforms can be used to invite the entire population of a basin to participate in consultations on a specific and well-defined issue or in events or actions which have an indirect impact on the basin's water resources management [27]. The use of digital platforms can increase access, integration and exploitation of environmental data and knowledge by scientific and end-user communities [15–17,19]. This is particularly relevant for cloud-based approaches where services (i.e., computing power and data storage) are also available to users which provide greater accessibility, efficiency and transparency to stakeholders.

The Basin Futures platform can be used to make scaling assessment areas and stakeholder engagement easier through user-friendly interfaces, intuitive workflow and the ability to move between multiple scales of observation (micro, meso and macro) and modelling within one platform. This can facilitate collaboration between landholders, scientists and policymakers through the transfer of knowledge and data from small-scale drainage basins to larger aggregates or transboundary basins, contributing to well-informed decisions and policies (Figure 3).

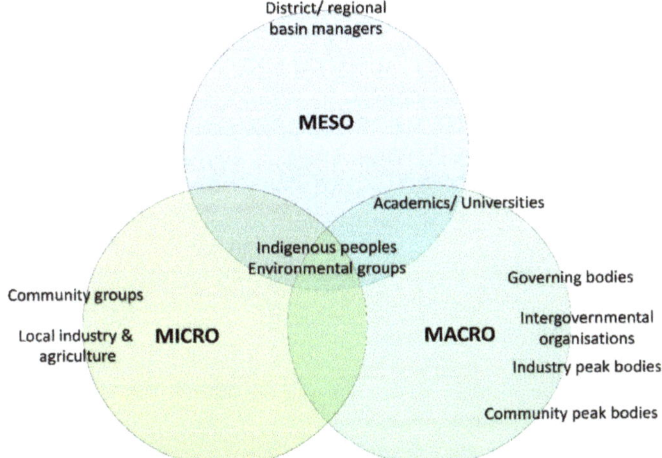

Figure 3. Representation of macro-, meso- and micro-level stakeholder groups.

4. Framework for Integrative Stakeholder Engagement

Despite the immense potential for digital platforms to be used for stakeholder engagement, there are limited guiding frameworks for how they should be implemented in conjunction with best practice IWRM principles. We present our framework for stakeholder engagement and collaboration on water-related issues and projects using digital platforms (Figure 4). The framework was generated based on IWRM principles and our collective experience of engaging with stakeholders on water resource developments and complex water management issues We then apply this framework to three case studies using the Basin Futures platform. The framework comprises of five main steps. The first is to identify the problem or issue at hand, for example, will the water supply be sufficient for future demands of a particular basin. The second step is to define the problem, identify the stakeholders, define goals and constraints and identify alternatives. The third step is the collaborate with stakeholders and explore options within the digital platform, including determining the current water balance and future changes through multiple scenarios. The fourth is to evaluate options explored and their environmental, economic and social impacts. The final step is to decide on an option or identify alternatives.

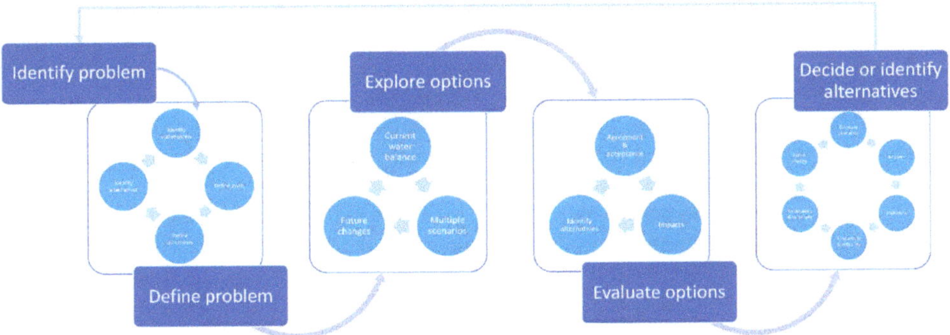

Figure 4. Framework for stakeholder engagement and collaboration on water-related issues and projects using digital platforms.

5. Case Study Applications of Basin Futures

5.1. Case Study 1: Stakeholder Engagement in Agricultural Development Projects

The world's population is projected to increase by 35% by 2050, which will require a 70–100% rise in food production given projected trends in diets, consumption, and income [6,28,29]. Over three-quarters of the projected increased population live in developing countries and in regions that already lack the capacity to produce enough food [30]. Increased food production can be achieved by increasing crop yields on existing farmland through better sustainable management practices or expanding crop production areas. Sustainable development of agricultural resources requires resolution and incorporation of diverse stakeholder values and interests. Collaboration and effective communication between local-scale farmers, governing bodies and research organisations can enable an efficient agriculture production system to be designed.

The Basin Futures platform can be used to explore and collaborate on agricultural development scenarios. The platform can be parameterised to incorporate current agriculture demands on an annual basis and forecast potential expansions, intensification and changes to agricultural systems as well as changes in water supply caused by other factors such as climate change. Basin Futures can assess the temporal reliability and production values of various agricultural scenarios and determine the impact on water resources and the environment. We selected the Purari Basin in Papua New Guinea (PNG) to demonstrate a potential stakeholder engagement strategy regarding agricultural development using the Basin Futures platform.

5.1.1. The Problem—Agricultural Development in the Purari Basin, Papua New Guinea

Agriculture plays a vital role in the Papua New Guinean economy. It employs ~50% of the national workforce, which generates 15% of gross domestic product (GDP) [31]. Despite its importance, agricultural productivity in PNG is generally low due to inadequate infrastructure and access to essential knowledge and farm inputs. It is estimated that 30% of the land is suitable for agriculture, yet only 2.2% is used commercially [31]. Papua New Guinea aims to enhance agricultural productivity, the scale of production, market access and income generation [31]. We use the Purari Basin as an example of a potential PNG basin to undergo agricultural development. The Purari Basin is located in the district of Chuave, Papua New Guinea. Purari Basin has an area of 33,080 km and an approximate population of 1.9 million people. The Basin experiences low water stress, but droughts and floods occur regularly.

5.1.2. Understanding the Problem

Papua New Guinea is a rural society with the majority of the population living in traditional communities [31]. There is limited commercial agriculture, and most produce is from small-scale, family-run farms. Small-scale farmers would provide invaluable insights into crop types, planting dates, expected yields and seasonal influences that would provide validation of the current agricultural baseline along with calibration of projected developments. The Purari Basin is home to several protected areas such as the Siwi-Utame Wildlife Management Area. Therefore, any major development is likely to require approval and engagement from indigenous peoples and national governing bodies such as the Conservation and Environment Protection Authority. Potential agricultural development would require a collaborative approach to stakeholder engagement across multiple assessment levels (Table 1).

Table 1. Stakeholder model inputs, contributions and outputs at various assessment levels within the Purari Basin. Directions of arrows indicate increases (↑) and decreases (↓) of outputs.

Stakeholders	Model Inputs	Assessment Level	Contribution	Outputs
Local farmers	Crop types and varieties, crop prices, planting date, maximum yields, cropping intensity, irrigated supply efficiency, soil characteristics, planting areas, fallow characteristics, suitability of area for agriculture	Micro	Local knowledge, empowerment, concerns and aspirations	↑ Knowledge ↑ Accuracy ↑ Participation ↑ Stakeholder empowerment and equity ↑ Likelihood of agreement ↓ Likelihood of conflict
Indigenous peoples	Suitability of area for agriculture	Micro/Meso/Macro	Local knowledge, cultural awareness	↑ Knowledge ↑ Awareness ↑ Inclusiveness ↑ Participation ↑ Stakeholder empowerment and equity ↑ Likelihood of agreement ↓ Likelihood of conflict
Environmental groups	Suitability of area for agriculture	Micro/Meso/Macro	Local/national knowledge, expertise, concerns	↑ Knowledge ↑ Awareness ↑ Inclusiveness ↑ Participation ↑ Stakeholder empowerment and equity ↑ Likelihood of agreement ↓ Likelihood of conflict
Academics	Crop types, crop prices, maximum yields, cropping intensity, irrigated supply efficiency, suitability of area for agriculture	Meso/Macro	Knowledge, evidence-based research	↑ Knowledge ↑ Accuracy ↑ Robustness
Local industry	Crop types, crop prices, maximum yields, cropping intensity, irrigated supply efficiency, suitability of area for agriculture	Micro/Meso	Local concerns, knowledge	↑ Knowledge ↑ Awareness ↑ Inclusiveness ↑ Participation ↑ Stakeholder empowerment and equity ↑ Likelihood of agreement ↓ Likelihood of conflict
States/district officials/regional basin managers	Current storages, crop types, maximum yields, cropping intensity, irrigated supply efficiency, planting areas, suitability of area for agriculture	Meso	Facilitating efforts, coordinating	↑ Awareness ↑ Inclusiveness ↑ Participation ↑ Stakeholder empowerment and equity ↑ Likelihood of agreement ↓ Likelihood of conflict

Table 1. *Cont.*

Stakeholders	Model Inputs	Assessment Level	Contribution	Outputs
Government bodies (National Department of Agriculture and Livestock)	Current storages, potential storages, crop prices, irrigated supply efficiency, planting areas, suitability of area for agriculture, nutritional values	Macro/Meso	Facilitating, funding, regulation	↑ Awareness ↑ Inclusiveness ↑ Participation ↑ Stakeholder empowerment and equity ↑ Likelihood of agreement ↓ Likelihood of conflict
Community peak bodies (i.e., Pacific Island Farmers Organisation Network)	Crop prices, planting areas, suitability of area for agriculture	Macro/Meso	Stakeholder representation, communication	↑ Awareness ↑ Likelihood of agreement ↓ Likelihood of conflict
Industry peak bodies	Crop prices, planting areas, suitability of area for agriculture	Macro/Meso	Regulation, communication	↑ Awareness ↑ Likelihood of agreement ↓ Likelihood of conflict
International agribusinesses	Crop prices, crop types	Macro	Global markets influence local farming practices (e.g., planting decisions)	↑ Awareness ↑ Likelihood of agreement ↓ Likelihood of conflict

5.1.3. Exploring and Evaluating Options with a Common Vision

Through the collaboration and use of the Basin Futures digital platform, a common vision for agricultural development could be attained by developing relevance and direction. Enabling cooperative, best-practice and efficient agriculture production systems to be designed. Channelling the efforts and knowledge of all stakeholders into the one platform can increase common understanding and agreements on priorities achieved through a fully inclusive, transparent and participatory process.

5.2. Case Study 2: Negotiating Water-Sharing Agreements

Water issues, hydrological boundaries and development impacts cut across administrative frontiers. Well-informed water management decisions and policies require effective communication and a shared understanding of the issue at hand between stakeholders. Communication and understanding between stakeholders can be difficult to achieve, especially in transboundary or politically contentious basins. Digital tools that are objective, transparent and cater to a range of abilities can pave the way forward to understanding issues and effective communication between stakeholder groups.

The purpose of Basin Futures is to digitally transform engagement processes and make accessible information and evidence for the majority, hence giving people voice and the ability to engage with the evidence for themselves and explore scenarios and their alternatives rather than just relying on more prescribed information. The use of digital platforms that can take into account global scale hydrological changes such as climate change and local scale water issues and developments and create tangible dialogues for action at appropriate national, regional, and local scales is a key to future stakeholder engagement processes. We selected the Mahanadi Basin to demonstrate a potential stakeholder engagement strategy for negotiating water-sharing agreements between the states of Chhattisgarh and Odisha using the Basin Futures platform.

5.2.1. The Problem—Water Sharing in the Mahanadi Basin, India

The Mahanadi Basin in India is home to ~36 million people. The Mahanadi River begins in the state of Chhattisgarh and flows through to the state of Odisha. The river is essential for intensive agriculture in the region, not only for irrigation but for extensive fertile soil the river deposits along the 858 km river course. The construction of the large Hirakud Dam has greatly altered the flow regime and extent of flooding that occurs in the region. Water sharing practices in the Mahanadi Basin have been a source of conflict between the Indian States of Chhattisgarh and Odisha. The two states are in dispute over the construction of allegedly illegal barrages within the respective states that are damaging the ecological integrity of the river. The states also disagree on the use of the Hirakud Dam and the catchment area share each state has of the reservoir.

5.2.2. Understanding the Problem

The Mahanadi Basin is essential for agriculture, industry, drinking water and the environment. Negotiating water-sharing agreements in this Basin would require an objective tool that would foster cooperation and partnership between contrasting stakeholder groups and governing bodies. The use of the Basin Futures platform could provide an easy exchange of information based on a common agenda across many production systems and ecological zones. Cooperation could be further facilitated through clear mechanisms and lines of communication. All efforts and resources of stakeholders should be channelled in that direction, guided by common understanding and agreements on priorities achieved through fully inclusive and participatory processes (Table 2). Furthermore, the use of the same datasets, methods and modelling platform would allow for unbiased dialogues and solutions to be created.

Table 2. Stakeholder model inputs, contributions and outputs at various assessment levels within the Mahanadi Basin. Directions of arrows indicate increases (↑) and decreases (↓) of outputs.

Stakeholders	Model Inputs	Assessment Level	Contribution	Outputs
Local farmers	Irrigated demands, timing of demands	Micro/Meso	Local knowledge, concerns, inputs	↑ Knowledge ↑ Participation ↑ Likelihood of agreement ↓ Likelihood of conflict
Indigenous peoples	Cultural values	Micro/Meso/ Macro	Local knowledge, cultural awareness	↑ Knowledge ↑ Participation ↑ Awareness ↑ Inclusiveness ↑ Likelihood of agreement ↓ Likelihood of conflict
Environmental groups	Environmental values and requirements	Micro/Meso/ Macro	Local/national knowledge, concerns, awareness	↑ Knowledge ↑ Participation ↑ Awareness ↑ Inclusiveness ↑ Likelihood of agreement ↓ Likelihood of conflict
Academics	Evidence-based water transfer partitioning	Meso/Macro	Knowledge, evidence-based research, range finding, system sensitivity	↑ Knowledge ↑ Accuracy ↑ Robustness ↑ Stakeholder empowerment and equity ↑ Likelihood of agreement ↓ Likelihood of conflict ↓ Unintended trade-offs in water benefits

Table 2. *Cont.*

Stakeholders	Model Inputs	Assessment Level	Contribution	Outputs
Local industry	Water demands and timing	Micro/Meso	Local concerns, knowledge, inputs	↑ Knowledge ↑ Participation ↑ Likelihood of agreement ↓ Likelihood of conflict
States/district officials/regional basin managers	Current demands, future demands, water allocation requirements	Meso	Facilitating efforts, coordinating, communication, awareness, prioritisation, range finding, alternative pathways, system sensitivity	↑ Awareness ↑ Inclusiveness ↑ Participation ↑ Stakeholder empowerment and equity ↑ Likelihood of agreement ↓ Likelihood of conflict ↓ Unintended trade-offs in water benefits
Government bodies	Current demands, future demands, water allocation requirements	Macro/Meso	Facilitating efforts, coordinating, communication, awareness, prioritisation, range finding, alternative pathways, system sensitivity	↑ Awareness ↑ Inclusiveness ↑ Participation ↑ Stakeholder empowerment and equity ↑ Likelihood of agreement ↓ Likelihood of conflict ↓ Unintended trade-offs in water benefits
Environmental peak bodies (i.e., World Wildlife Foundation)	Environmental values and requirements	Macro/Meso	Local/national knowledge, concerns, awareness	↑ Knowledge ↑ Robustness ↑ Accuracy ↑ Awareness ↑ Inclusiveness ↑ Participation ↑ Support ↑ Likelihood of agreement ↓ Likelihood of conflict
Industry peak bodies	Export demands	Macro/Meso	Regulation, communication	↑ Awareness ↑ Likelihood of agreement ↓ Likelihood of conflict
Intergovernmental organisations	Evidence-based water transfer partitioning	Macro	Regulation, communication	↑ Awareness ↑ Likelihood of agreement ↓ Likelihood of conflict

5.3. Case Study 3: Environmental Flows

Water resources produced by healthy ecosystems provide livelihood support for millions of people; this support is often extremely critical and essential for developing regions. River flow regimes are regarded to be the primary drivers of riverine and floodplain wetland ecosystems [32–34]. The flow regime is a major determinant of both biotic and abiotic components of a river system. Alteration of the natural flow regime can have serious consequences on the ecological sustainability of rivers and their associated floodplain wetlands [32,34]. The environment needs water to sustain itself, but in the water allocation decision-making process, the needs of the environment are often neglected [35]. If too much water is allocated to other sectors, the impacts on ecosystems can be devastating. A balance needs to be struck between people's direct water needs for domestic use, industry and agriculture and their indirect needs, through the numerous and unquantified goods and services provided by functioning ecosystems [36]. Stakeholder engagement and consensus in environmental flows can be difficult to obtain due to a lack of clarity on the issue, perceived non-transparent information and complexities of interacting perspectives.

The Basin Futures platform can be used to explore and collaborate on environmental flow scenarios. The platform is objective, transparent and allows users to run their own models, interpret results and form their own opinions instead of relying on more prescribed information. The platform can be used to range find and balance the inter-dependencies between hydrological, social, economic and ecological needs in river, lake and aquifer basins. We have selected the Pangani Basin in Tanzania, Africa, to demonstrate a potential stakeholder engagement strategy for environmental flows using the Basin Futures platform.

5.3.1. The Problem—Environmental Flows in the Pangani Basin, Tanzania

The Pangani River Basin in East Africa has a population of 2.6 million people. The Pangani River begins as a series of small streams near Mount Kilimanjaro and passes through the arid Masai Steppe before reaching its estuary and the Indian Ocean. Along its 500 km course, the Pangani River is a lifeline for biodiversity, people, and industry, and is fundamental to the economic development of the region. The Basin is home to Kilimanjaro National Park, a listed World Heritage Site with extensive biodiversity values. The Pangani Basin experiences medium to high water stress with a high flood occurrence. The Basin contains several critically endangered terrestrial and aquatic species.

5.3.2. Understanding the Problem

The Pangani Basin has widespread biodiversity values but is also used extensively for agriculture, industry and hydropower. Agreement on allocating water for the environment would have to include contrasting stakeholder groups with different perspectives and values. Consensus on how much, when and the variability of water provided to the environment would require a mechanism for negotiation using processes that enable the interests of traditionally more powerful water users and less powerful sectors to be reconciled [37,38]. Environmental flows must have clear objectives and scenarios built on multi-stakeholder consensus. Scientists can provide expert advice on how river basins change under various flow conditions, but it is the stakeholders who can say what the river is used for and how much water they need. The Basin Futures platform can be used to reconcile different stakeholder views and inputs to evaluate how ecology, economic costs and benefits across sectors and social equity respond to alternate river flow scenarios at multiple assessment levels (Table 3).

Table 3. Stakeholder model inputs, contributions and outputs at various assessment levels within the Pangani River Basin. Directions of arrows indicate increases (↑) and decreases (↓) of outputs.

Stakeholders	Model Inputs	Assessment Level	Contribution	Outputs
Local farmers	Irrigated demands, timing of demands	Micro/Meso	Local knowledge, concerns, inputs	↑ Knowledge ↑ Participation ↑ Likelihood of agreement ↓ Likelihood of conflict
Indigenous peoples	Environmental values, flow timing and requirements	Micro/Meso/ Macro	Local knowledge, cultural awareness, environmental and cultural inputs, prioritisation	↑ Knowledge ↑ Robustness ↑ Accuracy ↑ Awareness ↑ Inclusiveness ↑ Participation ↑ Stakeholder empowerment and equity ↑ Likelihood of agreement ↓ Likelihood of conflict

Table 3. *Cont.*

Stakeholders	Model Inputs	Assessment Level	Contribution	Outputs
Environmental groups	Environmental and cultural values, flow timing and requirements,	Micro/Meso/ Macro	Local/national knowledge, awareness, prioritisation, range finding, alternative pathways, system sensitivity	↑ Knowledge ↑ Robustness ↑ Accuracy ↑ Awareness ↑ Inclusiveness ↑ Participation ↑ Stakeholder empowerment and equity ↑ Likelihood of agreement ↓ Likelihood of conflict
Academics	Flow timing and requirements	Meso/Macro	Knowledge, evidence-based research, range finding, system sensitivity	↑ Knowledge ↑ Accuracy ↑ Robustness
Local industry	Water demands and timing	Micro/Meso	Local concerns, knowledge, inputs	↑ Knowledge ↑ Participation ↑ Likelihood of agreement ↓ Likelihood of conflict
States/district officials/regional basin managers	Current demands, future demands, water management strategies	Meso	Facilitating efforts, coordinating	↑ Awareness ↑ Inclusiveness ↑ Participation ↑ Stakeholder empowerment and equity ↑ Likelihood of agreement ↓ Likelihood of conflict ↓ Unintended trade-offs in water benefits
Government bodies	Current demands, future demands, water management strategies	Macro/Meso	Facilitating, funding, regulation	↑ Awareness ↑ Inclusiveness ↑ Participation ↑ Stakeholder empowerment and equity ↑ Likelihood of agreement ↓ Likelihood of conflict ↓ Unintended trade-offs in water benefits
Environmental peak bodies (i.e., World Wildlife Foundation)	Current demands, future demands, water management strategies, environmental and cultural values, flow timing and requirements	Macro/Meso	Stakeholder representation, communication, awareness, prioritisation, range finding, alternative pathways, system sensitivity	↑ Knowledge ↑ Robustness ↑ Accuracy ↑ Awareness ↑ Inclusiveness ↑ Participation ↑ Support ↑ Likelihood of agreement ↓ Likelihood of conflict
Industry peak bodies	Export demands	Macro/Meso	Regulation, communication	↑ Awareness ↑ Likelihood of agreement ↓ Likelihood of conflict

6. Assumptions and Disadvantages of Digital Platforms for Stakeholder Engagement

There are clear benefits to utilising digital platforms in the stakeholder engagement process. However, there are user assumptions and drawbacks of using technological-based approaches that must be considered. An assumption and disadvantage of using digital tools are that users are expected to be computer literate and have a working internet connection and access to electronic devices. While the world is becoming increasingly technologically connected, the adoption of digital platforms could further the equality gap between traditionally more powerful water users and less powerful sectors such as local farmers. These local farmers could be left out or disadvantaged by the engagement process if digital tools were solely used. Furthermore, in-person contact and interactions between stakeholders during the participatory process can reveal conflicts or issues that are otherwise not visible to water managers and planners. Loss of personal connections could increase the likelihood of conflict and disagreement of water management and development plans. Therefore, digital platforms should not be used as a complete replacement of traditional stakeholder engagement strategies but rather as a tool to compliment them.

7. Conclusions

Well-informed water management decisions and policies require effective communication and a shared understanding of the issue at hand between stakeholders. Communication and understanding between stakeholders can be difficult to achieve, especially in transboundary or politically contentious basins. Digital tools that are objective, transparent and cater to a range of abilities can pave the way forward to understanding issues and effective communication between stakeholder groups. Despite the volume of information, data and models that can be provided to stakeholders in regards to water-related issues, supported decisions can be difficult to obtain. In this paper, we demonstrated how our platform can be used to reconcile and channel differing stakeholder views into the one modelling platform by exploring three broad yet common water resource planning issues. Each case study explored interactions approximately between 10 stakeholders with interests at micro/meso or macro scale in the basin. The flexibility in transcending the scales with relative ease while ensuring accuracy and transparency leads to an increase in the overall participation and knowledge. While Basin Futures can be used to engage and collaborate with a variety of stakeholders on water-related issues and projects, it is essential that the process is ground-truthed with local data. The platform can be used to gain creative insights and a better understanding of the issue at hand to build consensus and agreement between stakeholder groups.

Author Contributions: Conceptualisation, J.O., C.P., A.P., A.S. and P.T.; methodology, J.O. and C.P.; software, P.T.; writing—original draft preparation, J.O.; writing—review and editing, J.O., C.P., A.P., A.S. and P.T. All authors have read and agreed to the published version of the manuscript.

Funding: This perspective piece contributes to the Basin Futures project, which is funded by CSIRO Land and Water.

Acknowledgments: We acknowledge the advisory board for Basin Futures. We thank three anonymous reviewers for improvements to the manuscript.

Conflicts of Interest: The authors declare no conflict of interest.

References

1. Vörösmarty, C.J.; McIntyre, P.B.; Gessner, M.O.; Dudgeon, D.; Prusevich, A.; Green, P.; Glidden, S.; Bunn, S.E.; Sullivan, C.A.; Liermann, C.R. Global threats to human water security and river biodiversity. *Nature* **2010**, *467*, 555. [CrossRef] [PubMed]
2. Mekonnen, M.M.; Hoekstra, A.Y. Four billion people facing severe water scarcity. *Sci. Adv.* **2016**, *2*, e1500323. [CrossRef] [PubMed]
3. UNESCO World Water Assessment Programme. *The United Nations World Water Development Report 2019: Leaving No One Behind*; UNESCO: Paris, France, 2019.

4. FAO. *Coping With Water Scarcity: An Action Framework for Agriculture and Food Security*; Food and Agriculture Organization of the United Nations: Rome, Italy, 2012.
5. Mancosu, N.; Snyder, R.; Kyriakakis, G.; Spano, D. Water scarcity and future challenges for food production. *Water* **2015**, *7*, 975–992. [CrossRef]
6. Rosegrant, M.W.; Ringler, C.; Zhu, T. Water for agriculture: Maintaining food security under growing scarcity. *Annu. Rev. Environ. Resour.* **2009**, *34*, 205–222. [CrossRef]
7. Montanari, A.; Young, G.; Savenije, H.; Hughes, D.; Wagener, T.; Ren, L.; Koutsoyiannis, D.; Cudennec, C.; Toth, E.; Grimaldi, S. "Panta Rhei—everything flows": Change in hydrology and society—The IAHS scientific decade 2013–2022. *Hydrol. Sci. J.* **2013**, *58*, 1256–1275. [CrossRef]
8. FAO. *Coping With Water Scarcity in a Changing Climate*; FAO: Rome, Italy, 2018.
9. Vörösmarty, C.J.; Green, P.; Salisbury, J.; Lammers, R.B. Global water resources: Vulnerability from climate change and population growth. *Science* **2000**, *289*, 284–288. [CrossRef] [PubMed]
10. Mackay, E.; Wilkinson, M.; Macleod, C.J.; Beven, K.; Percy, B.J.; Macklin, M.; Quinn, P.F.; Stutter, M.; Haygarth, P.M. Digital catchment observatories: A platform for engagement and knowledge exchange between catchment scientists, policy makers, and local communities. *Water Resour. Res.* **2015**, *51*, 4815–4822. [CrossRef]
11. Graversgaard, M.; Jacobsen, B.H.; Kjeldsen, C.; Dalgaard, T. Stakeholder Engagement and Knowledge Co-Creation in Water Planning: Can Public Participation Increase Cost-Effectiveness? *Water* **2017**, *9*, 191. [CrossRef]
12. Wehn, U.; Collins, K.; Anema, K.; Basco-Carrera, L.; Lerebours, A. Stakeholder engagement in water governance as social learning: Lessons from practice. *Water Int.* **2018**, *43*, 34–59. [CrossRef]
13. OECD. *Stakeholder Engagement for Inclusive Water Governance*; IWA Publishing: London, UK, 2015.
14. Verkerk, P.J.; Sánchez, A.; Libbrecht, S.; Broekman, A.; Bruggeman, A.; Daly-Hassen, H.; Giannakis, E.; Jebari, S.; Kok, K.; Krivograd Klemenčič, A.; et al. A Participatory Approach for Adapting River Basins to Climate Change. *Water* **2017**, *9*, 958. [CrossRef]
15. Olsson, J.A.; Andersson, L. Possibilities and problems with the use of models as a communication tool in water resource management. In *Integrated Assessment of Water Resources and Global Change*; Springer: Dordrecht, The Netherlands, 2006; pp. 97–110.
16. Faulkner, H.; Parker, D.; Green, C.; Beven, K. Developing a Translational Discourse to Communicate Uncertainty in Flood Risk between Science and the Practitioner. *AMBIO J. Hum. Environ.* **2007**, *36*, 692–704. [CrossRef]
17. Beven, K.J.; Alcock, R.E. Modelling everything everywhere: A new approach to decision-making for water management under uncertainty. *Freshw. Biol.* **2012**, *57*, 124–132. [CrossRef]
18. Hutton, C.; Wagener, T.; Freer, J.; Han, D.; Duffy, C.; Arheimer, B. Most computational hydrology is not reproducible, so is it really science? *Water Resour. Res.* **2016**, *52*, 7548–7555. [CrossRef]
19. Emmett, B.; Gurney, R.; McDonald, A.; Ball, L.; Bicak, M.; Blair, G.; Bloomfield, J.; Buytaert, W.; Delve, J.; Yehia, E.; et al. Heads in the cloud: Innovation in data and model dissemination. *Int. Innov.* **2014**, *141*, 82–85.
20. Cosgrove, W.J.; Loucks, D.P. Water management: Current and future challenges and research directions. *Water Resour. Res.* **2015**, *51*, 4823–4839. [CrossRef]
21. Taylor, P.; Stewart, J.; Rahman, J.; Parashar, A.; Pollino, C.; Podger, G. Basin Futures: Supporting water planning in data poor basins. In Proceedings of the 22nd International Congress on Modelling and Simulation, Hobart, Tasmania, Australia, 3–8 December 2017.
22. Perrin, C.; Michel, C.; Andréassian, V. Improvement of a parsimonious model for streamflow simulation. *J. Hydrol.* **2003**, *279*, 275–289. [CrossRef]
23. Frieler, K.; Lange, S.; Piontek, F.; Reyer, C.P.O.; Schewe, J.; Warszawski, L.; Zhao, F.; Chini, L.; Denvil, S.; Emanuel, K.; et al. Assessing the impacts of 1.5 °C global warming—Simulation protocol of the Inter-Sectoral Impact Model Intercomparison Project (ISIMIP2b). *Geosci. Model Dev.* **2017**, *10*, 4321–4345. [CrossRef]
24. Castrogiovanni, E.M.; La Loggia, G.; Noto, L.V. Design storm prediction and hydrologic modeling using a web-GIS approach on a free-software platform. *Atmos. Res.* **2005**, *77*, 367–377. [CrossRef]
25. Svatoň, V.; Podhoranyi, M.; Vavřík, R.; Veteška, P.; Szturcová, D.; Vojtek, D.; Martinovič, J.; Vondrák, V. Floreon+: A Web-Based Platform for Flood Prediction, Hydrologic Modelling and Dynamic Data Analysis. In *Dynamics in GIscience*; Springer: Cham, Switzerland, 2018; pp. 409–422.

26. McDonald, S.; Mohammed, I.N.; Bolten, J.D.; Pulla, S.; Meechaiya, C.; Markert, A.; Nelson, E.J.; Srinivasan, R.; Lakshmi, V. Web-based decision support system tools: The Soil and Water Assessment Tool Online visualization and analyses (SWATOnline) and NASA earth observation data downloading and reformatting tool (NASAaccess). *Environ. Model. Softw.* **2019**, *120*, 104499. [CrossRef]
27. International Network of Basin Organizations. *The Handbook for the Participation of Stakeholders and the Civil Society in the Basins of Rivers, Lakes and Aquifers*; INBO: Paris, France, 2018.
28. Bruinsma, J. The resource outlook to 2050: By how much do land, water and crop yields need to increase by 2050. In Proceedings of the Technical Meeting of Experts on How to Feed the World in 2050, Rome, Italy, 24–26 June 2009; pp. 24–26.
29. Van Wart, J.; Kersebaum, K.C.; Peng, S.; Milner, M.; Cassman, K.G. Estimating crop yield potential at regional to national scales. *Field Crops Res.* **2013**, *143*, 34–43. [CrossRef]
30. Tilman, D.; Balzer, C.; Hill, J.; Befort, B.L. Global food demand and the sustainable intensification of agriculture. *Proc. Natl. Acad. Sci. USA* **2011**, *108*, 20260–20264. [CrossRef] [PubMed]
31. Ministry of Agriculture and Livestock. *National Agriculture Development Plan 2007–2016*; Ministry of Agriculture and Livestock: Port Moresby, Papua New Guinea, 2007.
32. Bunn, S.E.; Arthington, A.H. Basic principles and ecological consequences of altered flow regimes for aquatic biodiversity. *Environ. Manag.* **2002**, *30*, 492–507. [CrossRef] [PubMed]
33. Junk, W.J.; Bayley, P.B.; Sparks, R.E. The flood pulse concept in river-floodplain systems. *Can. Spec. Publ. Fish. Aquat. Sci.* **1989**, *106*, 110–127.
34. Poff, N.L.; Zimmerman, J.K. Ecological responses to altered flow regimes: A literature review to inform the science and management of environmental flows. *Freshw. Biol.* **2010**, *55*, 194–205. [CrossRef]
35. Leendertse, K.; Mitchell, S.; Harlin, J. IWRM and the environment: A view on their interaction and examples where IWRM led to better environmental management in developing countries. *Water SA* **2008**, *34*, 691–698. [CrossRef]
36. Sullivan, C. Calculating a Water Poverty Index. *World Dev.* **2002**, *30*, 1195–1210. [CrossRef]
37. Norman, A. Water resources sector strategy: Strategic directions for World Bank engagement. *J. Econ. Lit.* **2004**, *42*, 981.
38. Butler, C.; Adamowski, J. Empowering marginalized communities in water resources management: Addressing inequitable practices in Participatory Model Building. *J. Environ. Manag.* **2015**, *153*, 153–162. [CrossRef]

Article

Re-Interpreting Cooperation in Transboundary Waters: Bringing Experiences from the Brahmaputra Basin

Anamika Barua [1], Arundhati Deka [1,*], Vishaka Gulati [1], Sumit Vij [2], Xiawei Liao [3] and Halla Maher Qaddumi [4]

1 Department of Humanities and Social Sciences, Indian Institute of Technology Guwahati (IITG), Guwahati 781039, Indian; anamika.barua@gmail.com (A.B.); vishaka.gulati@gmail.com (V.G.)
2 Public Administration and Policy Group, Wageningen University and Research, 6706 KN Wageningen, The Netherlands; sumit.vij@wur.nl
3 School of Environment and Energy, Peking University Shenzhen Graduate School, Shenzhen 518055, China; xliao@worldbank.org
4 The World Bank, Washington, DC 20433, USA; hqaddumi@worldbank.org
* Correspondence: arundhatideka91@gmail.com; Tel.: +91-9811468460

Received: 15 October 2019; Accepted: 5 December 2019; Published: 8 December 2019

Abstract: Several studies have demonstrated the continuum of cooperation on transboundary rivers, but have largely focused on government to government (Track 1) cooperation and formal diplomacy. Formal arrangements like treaties, agreements, joint mechanisms, joint bodies, joint commissions (e.g., river basin organizations), etc., fall within the scope of transboundary waters cooperation. However, in some transboundary rivers, often due to political constraints, Track 1 cooperation might not be a feasible option. When governmental cooperation is a non-starter, effort and progress made outside the government domain through informal dialogues can play a significant role. It is therefore important to re-examine the definition of cooperation as it applies to international rivers, and potentially to broaden its scope. Such an examination raises important questions: What does international cooperation in this context actually mean? Is it formal (Track 1) cooperation related to sharing of water, data, and information only, or does it have a broader meaning? What, precisely, can be the entry point for such cooperation? Are informal transboundary dialogues and water diplomacy itself an entry point for cooperation on international rivers? This paper aimed to answer these critical questions drawing from the "Brahmaputra Dialogue" project initiated in 2013 under the South Asia Water Initiative (SAWI), which involved the four riparian countries of the Brahmaputra Basin. Several important focal points of cooperation emerged through this sustained dialogue, which went beyond sharing hydrological data or signing a basin-level treaty, broadening the definition of "cooperation". The paper, bringing evidence from the dialogue, argues that the Brahmaputra Dialogue process has led to a broader understanding of cooperation among basin stakeholders, which could influence water resource management of the basin in the future.

Keywords: transboundary waters; cooperation; integrated water resource management; Brahmaputra River Basin; South Asia

---

1. Introduction

The 2030 Agenda for Sustainable Development and its Sustainable Development Goals (SDG) framework considers transboundary water cooperation critical to development, prosperity, and peace. Target 6.5 of SDG 6 (Ensure access to water and sanitation for all), in particular, emphasizes the need to implement integrated water resources management (IWRM) and the need to include a transboundary dimension. Transboundary water cooperation is thus crucial to fully achieving the 2030 Agenda for Sustainable Development.

"Cooperation" is defined here as coordination between states, where they collaborate to achieve common interests with mutual benefits. To promote greater cooperation around the world's international river basins, significant efforts have been underway in the decades since the Dublin and Rio conferences. (In 1992 the International Conference on Water and the Environment was held in Dublin, Ireland. The output from this conference was a declaration regarding water that was presented to the United Nations Conference on Environment and Development (UNCED) that was held in Rio de Janeiro in June that year where the ideas from the 1987 UN Report (the Brundtland Report) were developed and discussed. The Rio conference, which came to be known as the "Earth Summit", was attended by 118 heads of government and was a major turning point in bringing the issues of sustainability and sustainable development onto the international political stage. The inclusion of the Dublin Principles in the conference debate helped to highlight the importance of water as a resource for environmental protection and human development.) States now also have general frameworks under international law applicable to the non-navigational uses of transboundary rivers and lakes in the form of the United Nations Watercourses Convention on the Law of the Non-Navigational Uses of International Watercourses (UNWC) and the UNECE Convention on the Protection and Use of Transboundary Watercourses and International Lakes. The water conventions provide riparian countries with a framework for cooperation, a common language and platform upon which States can negotiate equitable and sustainable solutions [1], identify common interests, and develop actions toward mutual benefits.

There is also a growing body of literature highlighting that conflict and cooperation can co-exist in transboundary waters situations [2–6]. There are several river basins across the world, such as the Nile, Jordan, and Mekong, which demonstrate that water, by its very nature, tends to induce even non-cooperative co-riparians to cooperate. While water has the ability to pose a threat (with scarcity leading to competition for the resource), it can also provide opportunity for increasing cooperation. The continuum of such water cooperation, as demonstrated by different studies, is mostly related to the direct mutual benefits from the water resources, such as coordinated water management plans, hydrological data exchange, joint water infrastructure development, flood management, etc. As such, formal arrangements like treaties, agreements, joint mechanisms, joint bodies, joint commissions (e.g., river basin organizations), etc., fall within the scope of transboundary cooperation. The water conventions also support the development of such agreements, the establishment of joint bodies, and strengthening of institutions through the implementation of basin-level projects. Thus, the focus of "transboundary cooperation" has largely been on government-to-government (Track 1) cooperation and formal diplomacy; such cooperation is driven by the political moods of the riparian countries and is mostly negotiated through an official process of transboundary interactions, making cooperation over transboundary rivers a complex and inherently political process [7,8]. The different tracks of diplomacy can be defined as:

- Track 1 (traditional official diplomacy): Dialogues or negotiation between officials, which mostly include politicians, policy makers, and high-ranking military personnel in a nation-state centered perspective.
- Track 1.5: "Diplomatic initiatives that are facilitated by unofficial bodies, but directly involve officials from the conflict in question" [9].
- Track 2: As defined by [10], "unofficial, informal interaction between members of adversary groups or nations, who can interact more freely than high-ranking officials, to develop strategies, to influence public opinion, and organize human and material resources in ways that might help resolve their conflict".
- Track 3: People-to-people or grassroots-level diplomacy undertaken by individuals, civil society, and private groups to encourage interaction and understanding of communities' issues, and to generate awareness for empowerment within these communities [11].

Water diplomacy facilitates communication between sovereign states with the aim of promoting constructive cooperation and preventing conflicts over shared water resources [12]. While traditionally, diplomacy is defined as high-level interaction and dialogue between nation-states, in the present context, the definition has been broadened to include various other levels as well [13]. Hence, in the transboundary context, Track 1.5 and Track 2 diplomacy has played a significant role in several river basins in building trust and confidence of multiple stakeholders. Such efforts and progress being made outside of the government domain through informal diplomacy can play a significant role when governmental cooperation is a non-starter [2]. However, cooperation achieved through such informal diplomacy has usually remained outside the scope of "transboundary cooperation", because this cooperation is not directly related to benefits from the water resources.

This cooperation mostly takes the form of civil society collaborations on transboundary concerns, joint research undertaken by academics for knowledge creation, and joint stories developed by media personnel for the river basin, etc. In order to encourage such endeavors dedicated to building trust between multiple stakeholders sharing the same rivers, and to create socio-political environments that enable potential "formal" cooperation, it is therefore important to re-examine the definition of cooperation around transboundary rivers and potentially broaden its scope. Such a re-examination raises a few pertinent questions, such as: What does transboundary cooperation actually means? Is it formal (Track 1) cooperation related to sharing of water, data, and information only, or does it have a broader meaning? What precisely can be the entry point of such cooperation? Are informal transboundary dialogues and diplomacy itself an entry point of cooperation on international rivers?

The paper aimed to answer these critical questions, drawing from the "Brahmaputra Dialogue" (Transboundary Policy Dialogue for Improved Water Governance of Brahmaputra River) project initiated in 2013 under the South Asian Water Initiative (SAWI), which involves the four riparian countries of the Yarlung–Zangbo–Brahmaputra–Jamuna River Basin (herein referred to as Brahmaputra Basin). The Brahmaputra Dialogue is an informal platform, and was initiated to assist communication at different tracks and between different actors (representatives of states, civil society, academia, etc.) across the basin countries, in establishing connections and building trust. Significant avenues of cooperation emerged due to a sustained dialogue that went beyond sharing hydrological data or signing a basin-level treaty, broadening the definition of "cooperation". The paper, bringing evidence from the dialogue, discusses how the Brahmaputra Dialogue process has led to a broader understanding of "cooperation" among basin stakeholders, which could influence water resource management of the basin in the future.

This article is divided into four additional sections. The second section (next) brings in a conceptual discussion of transboundary cooperation. The third section explains the methodology used for data collection and analysis. It also briefly sets the context of the Brahmaputra River Basin. Section 4 presents the findings of this article, explaining how different elements of cooperation are emerging via the Brahmaputra Dialogue. The last section presents a discussion and concludes the article.

2. Conceptual Discussion on Transboundary Cooperation

Transboundary cooperation has numerous challenges, as the potential and incentive for each sovereign state to cooperate varies [11]. Cooperation requires an understanding to be formed of the diverse interests of stakeholders with respect to water resources, ensuring the sustainable development of a river or lake basin as a whole [14]. Integrated water resources management (IWRM) and transboundary water management are therefore two important components of SDG 6 (Target 6.5), and are intrinsically connected to the other principles of the SDGs and their targets. Progress towards SDG Target 6.5 is monitored through two indicators: 6.5.1 tracks the degree of implementation of IWRM at all levels, and Indicator 6.5.2, specific to transboundary water cooperation, is defined as the "proportion of transboundary basin area with an operational arrangement for water cooperation". These indicators were agreed by the United Nations Statistical Commission in March 2016 and were subsequently adopted in July 2017 by the United Nations General Assembly as part of the global indicator framework for the Sustainable Development Goals and Targets of the 2030 Agenda for Sustainable Development.

Transboundary Rivers, by their basic nature of crossing one or more political boundaries, are in the realm of international relations engagement between two or more nation states. International relations (IR) theory can help us understand the way the international systems work, as well as how nations engage with each other and view the world [15]. Various schools of thought in international relations—realists, liberal institutionalists, and constructivists—have theories on conflict and cooperation. The realists concentrate on hard military power and why cooperation is very difficult and complicated to achieve among states [16]. They state that all nations are working to increase their own power, and that those countries that manage to horde power most efficiently will thrive, as they can easily eclipse the achievements of less powerful nations. The liberals, also called "liberal internationalism", believe that the current global system is capable of engendering a peaceful world order. They believe in the power of institutions, and rather than relying on direct force, such as military action, liberalism places an emphasis on international cooperation as a means of furthering each nation's respective interests [17,18]. Constructivists rest on the notion that rather than the outright pursuit of material interests, it is a nation's belief systems—historical, cultural, and social—that explain its foreign policy efforts and behavior [19,20]. Constructivists also argue that states are not the most important actors in international relations, but that international institutions and other non-state actors are valuable in influencing behavior through lobbying and acts of persuasion [19]. [21], however, describes the framing of conflict and cooperation within the mainstream IR theory (realist and new liberalism) as a binary pair, which has led to the assumption that conflict and cooperation are the two basic or ideal types of international interaction. The overwhelming use of mainstream approaches, as Selby argues, has led to a narrow understanding of conflict and cooperation, and "cooperation" is invariably defined in opposition to "conflict". Due to such binary framing within mainstream IR theory, there is a strong value judgment, even within water-specific literature, that "cooperation is good and conflict is bad" [21]. As such, cooperation over water is considered to be the policy goal, irrespective of how it is achieved and who gains from such cooperation.

Within water literature, cooperation over transboundary water has evolved over the years. In the 1990's, water was looked at more from a conflict lens, which gave rise to a "Water Wars Thesis". The Water War Thesis argued that an inevitable global water crisis is advancing, which will trigger international wars between states [22,23] or two or more countries solely over water. The thesis drew support from water resource development literature as well as from the international rivers literature, which also focused on the possibility of disputes over water spilling over into outright conflict between states. However, the water war thesis has become the subject of extensive critique in recent years. For example, [24–26] have argued that disputes over water very rarely develop into acute militarized conflicts, as this would jeopardize the use of the resource itself. The critics even argued that water conflicts have actually encouraged cooperation between states. [27] claimed that, historically, cooperative efforts have always overpowered violent disputes over transboundary waters. There is a consensus among water professionals that the cooperative management of shared river basins should provide opportunities to increase the scope and scale of benefits [28,29]. In fact, [6,30] went a step further and even rejected the mainstream IR focus on conflict and cooperation in favor of a broader analysis of relations of power and hegemony within transboundary basins.

The hydro-hegemony analysts of international water politics (see [6,30]) have provided a broader analysis of relations of power and hegemony within transboundary basins. They contend that to understand the water issues in any river basin it is important to understand the politics involving the nation states in that particular river basin [6]. [4] took this research a step forward and emphasized that, in the vast majority of hydro-political contexts, conflict and cooperation co-exist. There is also recognition that power relations are asymmetrical, particularly between upstream and downstream countries, and that "not all cooperation is pretty" [4]. Hence, the conceptual frameworks that argue that conflict and cooperation can exist simultaneously in any river basin, without reverting to a "water wars" scenario [31], provide a counter-narrative to the assumptions that have held conflict and cooperation as essentially opposite ends of the spectrum of interactions. These researchers also put forth that cooperation should not be looked at as the end product of any international negotiation or international legal principle, as gains from cooperation may be unevenly distributed [21].

While these insights are important and have helped to move thinking away from the binary understanding of conflict and cooperation, in the view of the authors of this paper, the definition of cooperation still needs further refinement. The transboundary interactions that this water literature has focused on or analyzed are confined mostly to the state actors. Concepts like hegemony, power, power asymmetries, and domination etc., which have been prefixed to cooperation, mostly describe the engagement of state actors (between upstream and downstream countries) in international negotiations or interactions (i.e., Track 1 diplomacy leading to river basin organizations or institutional arrangements). At international levels, legal and institutional frameworks also center around such normative emphasis on state-driven cooperation. For example, the Convention on the Protection and Use of Transboundary Watercourses and International Lakes (Water Convention) provides a key legal and intergovernmental framework for promoting transboundary water cooperation [32]. The framework fosters the IWRM approach, and emphasizes that parties bordering the same transboundary waters should cooperate by entering into specific agreements and establishing joint bodies. An interstate agreement, or a joint body, joint mechanism, or commission that commonly governs transboundary rivers, such as a river basin organization, is considered transboundary cooperation. These interstate agreements often incorporate water convention principles such as "equitable and reasonable utilization" and "sustainable development" of shared watercourses with "no significant harm".

While treaties or agreements are important to bringing stability and enhance security in a transboundary river context, establishing and ensuring such long-term cooperation at the transboundary scale requires strong political commitments from the riparian countries. Generating such political will is not only challenging, but is also a drawn-out process. As politics take center stage, transboundary cooperation becomes complex and extremely challenging. There is always a risk of change in political leadership, and efforts made to generate willingness at the political level (Track 1) may not lead to fruition. Furthermore, as discussed above, in most cases such state-driven cooperation is skewed and benefits only the powerful riparians. Cooperation through asymmetrical treaties (like for the Nile, Jordan, and Ganges) has become a source of conflict rather than cooperation [33]. Therefore, [21] poses an interesting question: "Do there exist, or should there exist, limits to the idea of international 'cooperation'?"

In this paper, we focused on this aspect of cooperation and have argued that there is a need to extend the focus of transboundary cooperation beyond state actors. Cooperation can happen at multiple levels and between multiple stakeholders. There is a need to move from a purely analytical perspective, primarily centered on the role of the state [2,34–38], and to include the influence that non-state actors have on managing the river. Focusing only on state-driven cooperation denies or undermines cooperation that is driven by non-state actors at multiple levels, and it narrows the scope of the definition of cooperation to only what is defined in legal frameworks.

There are already several initiatives underway in river basins where either cooperation has been a non-starter or a treaty has led to conflict among the countries signing the agreement. One example is the Indus Basin Knowledge Forum, which helps to connect multiple stakeholders of the Indus Basin, shared by Afghanistan, China, India, and Pakistan. Some 300 million people live within the basin and rely on its resource base, and many more benefit from the harnessing of the basin's resources. India and Pakistan, the countries with the most area within the basin, divided up rights to the various tributaries under the Indus Water Treaty of 1960 (IWT). The IWT has survived various wars and other hostilities between the two countries, and, as such, it is largely considered a success. Today, however, the treaty is increasingly facing challenges it was not designed to address [39]. Growing demand for water and energy in both India and Pakistan, coupled with uncertain climate futures, has put the treaty under increasing stress, leading to a complex decision-making environment. Despite significant expertise and donor support over several decades, water management across the Indus Basin remains poor. Poor water resources management could be viewed as a missed opportunity to drive resilient economic growth and poverty reduction, while vulnerability to floods and droughts remains prevalent. Against this backdrop, the Indus Basin Knowledge Network (IBKN) or Indus Basin Dialogue was initiated in June 2013—an informal mechanism comprising participants from each of the four basin countries. The network has been able to bring together a wide range of stakeholders (including policy makers, development practitioners, academics, civil society organizations, and media), to increase the likelihood of information exchanges, which, in turn, could inform change in the basin. The dialogue has been able to build trust among stakeholders across the riparian countries through an Indus-Basin-wide dialogue process, providing capacity building, generating knowledge and information sharing, assessing climate change impacts, and promoting data exchange and collaborative research. This progress is significant, as it is a Track 2 dialogue involving participants from all four countries (not just India and Pakistan). The process is helping to build an enabling environment for cross-border collaboration on research as well as to ensure longer-term sustainability for the dialogue.

Similarly, IHE Delft has initiated new research in the Nile Basin on the role that journalists and scientists can play in transboundary conflicts or cooperation. The Nile—particularly the Blue Nile shared by Egypt, Sudan, and Ethiopia—is one of the international rivers often described as being on the verge of a "water war", as a consequence of competing claims and concurrent projects of water exploitation by the riparian countries. The "Open Water Diplomacy: Media, Science and Transboundary Cooperation in the Nile Basin" project aims to offer a space where water journalists and water scientists from different Nile Basin countries can get acquainted and engage in a process of common learning and co-production of knowledge. The project also aims to reach out to water diplomats—national governments, international institutions, NGOs involved in transboundary water management—to contribute to building shared narratives and a culture of cooperation in the Nile Basin.

In the Lancang–Mekong River—which originates from China as Lancang River and flows through Myanmar, Lao, Thailand, Cambodia, and Vietnam as Mekong River—China, as the first upstream country, is at a strategic geopolitical position and of paramount importance in terms of transboundary river cooperation in this region. While China's first multi-lateral engagement began with becoming an observer of the Mekong River Commission (MRC) in 1996—a regional mechanism founded by Laos, Thailand, Cambodia, and Vietnam—its most influential and recent engagement is probably the initiation of the Lancang–Mekong Cooperation (LMC) mechanism, including all six countries, in 2015. However, it is often overlooked that Chinese engagement in this region has been shaped outside the governmental domain for a very long time, both before and after the LMC was launched. For instance, in 2008, the Department of International Relationship at Yunnan University held an international academic workshop on the "Greater Mekong Subregion Economic Corridor Construction: Cooperation and Development", which was attended by more than 70 academics from Cambodia, Laos, Myanmar, Thailand, Vietnam, India, Japan, South Korea, and China. In 2019, the Lancang–Mekong Youth Exchange and Cooperation Center was established jointly by six universities from all six countries at Fudan University in Shanghai, with a memorandum of understanding signed to promote communication among youths in this region. Such collaboration and communication among the academic communities is playing an important role in creating a cooperative atmosphere and momentum, especially among the citizens, for encouraging any official cooperation among governmental entities.

Initiatives such as those discussed above can make a substantial contribution, particularly when Track 1 cooperation is challenging due to political constraints. A significant effort is needed to strengthen transboundary water cooperation and to realize its potential to support SDG6 and the many other water-related SDGs. Effort and progress made outside the government domain, through informal dialogues, can play an important role when governmental cooperation is a non-starter. These endeavors are dedicated to building trust between nations sharing the same rivers and creating enabling socio-political environments, as potential 'formal' cooperation needs to be encouraged. Hence, this paper argues that it is important to re-examine the definition of cooperation on international rivers and to potentially broaden its scope.

3. Methodology

3.1. Physical and Political Context of the Brahmaputra Basin

The Brahmaputra Basin (please see Figure 1) originates in the Tibet Autonomous Region (TAR) of China, has a basin drainage area of 580,000 km^2, and empties into the Bay of Bengal [40]. It is shared between four countries—China (50.5% of the total basin area), India (33.6%), Bangladesh (8.1%), and Bhutan (7.8%). The climate of the basin is monsoon (south-west) driven, with a distinct monsoon season from May to September accounting for 60–70% of the annual rainfall, but the upper flow is supported by groundwater and glacial/snow melt. The annual flow of the Brahmaputra River from China to India is estimated to be 165.4 billion cubic meters (BCM), with an additional 78 BCM entering India from Bhutan. As the river descends from Tibet, increased precipitation supports the growth of forests such as sal, a valuable timber tree found in Assam. At lower elevations, tall reed jungles grow in the swamps and depressed, water-filled areas (jheels) of the floodplains. Communities in the Assam Valley primarily grow tea in the upstream region, and cultivate fruit trees including plantains, papayas, mangos, and jackfruit. In this region, one can find 220 languages originating from three distinct language families—Indo-Aryan, Sino-Tibetan, and Austric. "The Brahmaputra basin lies in distinct geological and climatic zones, extending from the dry region of Tibet in the rain shadow of the Himalaya to the eastern basin receiving extremely high rainfall" [40,41].

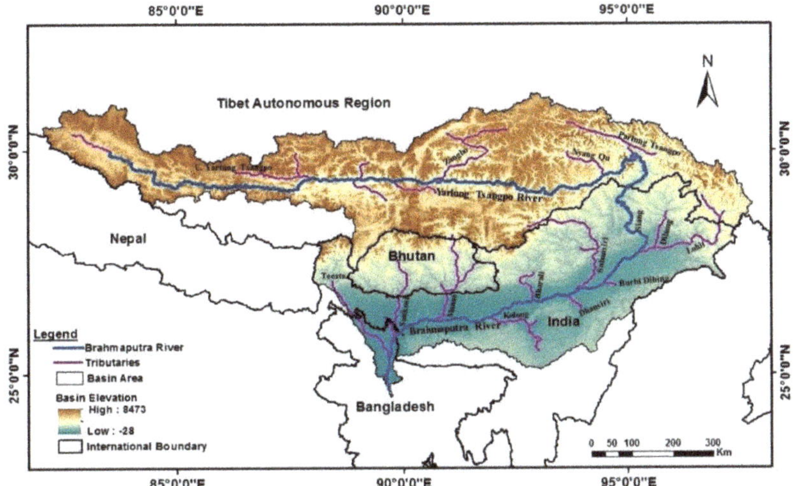

Figure 1. Map of Brahmaputra Basin (Source: IITG).

The basin has a varied terrain, high seasonal variability of river flow, and is also susceptible to sudden channel migration, making it a highly unpredictable and complex water system [40]. More than 100 million people live in the basin and the economic structure is highly river-dependent, with livelihoods relying on agriculture, livestock, forestry, and fisheries, among others. While the river is the primary source for the basin communities, a majority of the communities are marginalized and live in poverty. Although the river basin has immense potential to reduce poverty, with opportunities for irrigation development, livelihood enhancement, and operations such as inland water navigation and hydropower development, they have not been well harnessed [11].

Water scarcity is severe in South Asia and among other regions globally, and it is expected to get worse in the coming time. The hydrological impact of climate change on Brahmaputra Basin is expected to be greater than that of other basins, as it will be contributed to by glacial melt and extreme monsoon rainfall [42]. Monsoon is characterized by seasonal variations in rainfall in the region and the streamflow is likely to be affected due to climate change, with an increase in rainfall during wet summer period [43], but less rainfall during the dry winter period [44]. Over the downstream of the basin, seasonal fluctuations of surface water availability and water demand are out of phase (inflow of a large volume of surface water is limited in a relatively short monsoon season). During the dry season, there is a serious water shortage, with water demand exceeding water availability [45]. The population, along with their food demand and economic development of the basin, is anticipated to rise at a faster rate compared to other regions [44].

The basin is rich in biodiversity [46–48], but, the riparian countries face challenges related to floods and droughts, development of infrastructure, and lack of open communication, both within and between the countries. Consequentially, within the Brahmaputra Basin, there are the stereotypical conflicts of interest between upstream and downstream riparians, related to water resources development and water diversion plans of the upstream areas [46]. As each riparian country has a national priority with regard to the Brahmaputra, the understanding of benefit from the Brahmaputra River, therefore, varies between the four sovereign states of the basin, along with the incentives to cooperate. While the river provides economic and energy opportunities for China, India's main concerns include control of floods and erosion and harnessing the river's potential (through the development of hydroelectricity and navigation) to foster integration of North-East India (which is relatively isolated) with the rest of the country. For Bangladesh, it is crucial to manage the physical impacts of the river (like riverbank erosion, annual flooding, sedimentation, and diminished water flow in the dry season) [49,50]. Being the most downstream country, Bangladesh sees the development of water infrastructure in India and China as a threat. Conflicts often arise between India and Bangladesh regarding the strategies being adopted for controlling floods and harnessing of the potential of the Brahmaputra [7]. Each country does realize the potential the river provides for economic development, but the benefits are seen through localized and sectoral lenses, which trigger tension and disputes within, as well as between, the riparian countries. While the three riparian countries have not been able to harness the potential of the river together, India and Bhutan have been able to achieve some cooperation through the development of hydropower projects [11].

Other key concerns and challenges that are typical to the Brahmaputra Basin countries are historical rivalries (China–India war of 1962 and their border disputes), high political mistrust and suspicion, increasing nationalism, closed-door negotiations exclusively on water issues, and absence of negotiation frameworks [47]. Unlike in other international river basins, there is no institutional mechanism in place to address the issue of water management at the river basin level [48]. There are few bilateral agreements (Memoranda of Understanding) between the riparian countries addressing water-related issues like data sharing and flood forecasting. The overall scope of cooperation through such avenues is quite narrow. While the lower riparian countries insist on getting continuous data and information, it is shared only during the wet season [46]. To date, no multi-lateral or basin-wide agreement has been signed regarding the Brahmaputra Basin.

There are ongoing discussions among the political leaders for regional multi-lateral cooperation on water management of the Brahmaputra Basin, but very little progress has actually been made in achieving cooperation at the Track 1 level (i.e., government-to-government). For example, there were plans of instituting a Brahmaputra Valley Authority within India, similar to the Tennessee Valley Authority, but it never materialized [46]. Further, there is a lack of scientific knowledge and information about the river, as the Brahmaputra River Basin is relatively under-researched compared to other river basins in South Asia. This lack of information has not hampered the construction of water infrastructure projects on the river, especially by India and China. However, very little information about these projects is made available in the public domain, which has created mistrust and suspicion among the riparian countries [11,46].

A vital factor in the case of Brahmaputra Basin is the lack of a reliable and comprehensive network of basin-wide information on climate change, flow data, natural hazards, and economic factors (agricultural production, prices, and trade through navigation) [40]. To reduce the pressure on water demand due to the region's growing population and high development activities, long-term sustainable planning (population control, land use policy) is required. Some other non-structural measures that are significant for reducing exposure and social vulnerability could include development of an early warning system and implementation of water policy that benefit the marginalized. To mitigate the risk of water scarcity and to secure the livelihood of the communities, adaptation strategies need to be jointly discussed by the policy makers, researchers and grassroots level stakeholders across the countries, and river basin management authority of the region would require consolidation of relevant institutional mechanisms at various governance scales [51].

Water being both a center and state subject in India along with the central level institutions, it is the state-level institutions of Assam and Arunachal Pradesh that are involved with the river. Within the country, it is the states, not the central government, that have primary jurisdiction over the management of water resources. In the case of China, water is a national property and is therefore administered by the national-level ministry, i.e., MWR (Ministry of Water Resources of the People's Republic of China), which has the power to formulate sector policies, regulations, and laws. However, policy implementation and enforcement fall on the shoulders of provincial water bureaus, who are supposed to obey both of their superiors, i.e., MWR and provincial government. In Bangladesh and Bhutan, the river is managed primarily by the national-level government. Institutions present at the local level are involved only during implementation and consultation during planning of activities. Meanwhile, the transnational aspect pertaining to this basin, is missing from all the countries with the absence of any regional-level authority.

In order to develop trust and confidence between the riparian countries of the Brahmaputra and to work on the aforementioned issues and potential development agendas, there has to be long-term interaction and communication between different stakeholders, which should also include non-traditional stakeholders, such as the private sector, media, funding institutions, and marginalized groups, including women. Such multi-track diplomacy for the Brahmaputra Basin will create and support spaces where meaningful conversation can take place among diverse stakeholder groups. Such interaction can eventually inform and help shape more formal negotiations and decision making [52].

With this backdrop, in 2013, a multi-lateral and multi-track dialogue was initiated by SaciWATERs (a non-governmental organization based in India) (South Asia Consortium for Interdisciplinary Water Resources Studies) for the Brahmaputra Basin, with the aim of enhancing the interaction between multiple stakeholders. The dialogue initiated by SaciWATERs is, to date, the only multi-track and multi-lateral initiative that involves all the four basin countries and deals with the Brahmaputra River Basin. The first phase of the dialogue was supported by The Asia Foundation, and from 2014 onward, the dialogue became part of World Bank's SAWI project. In this paper, discussion is concentrated on the progress of the dialogue under the SAWI initiative. The dialogue has recently become institutionalized, with government funded research/academic institutes becoming the nodal partners (India: Indian Institute of Technology, Guwahati (IITG) and the regional nodal institute, Bangladesh: Institute of Water Modelling (IWM)) in each riparian nation for facilitating the dialogue.

3.2. Data Collection

For the purpose of this research, we studied the Brahmaputra Dialogue meetings in 2014–2018. The dialogue took a constructivist approach, as it is believed that both state and non-state actors are important stakeholders in transboundary water management, and that non-state actors can make a valuable contribution to paving the path of cooperation between state actors. We collected all the reports of the Brahmaputra Dialogue (BD) organized between 2014 and 2018 (see Appendix A for more details). In three phases, 23 workshops and meetings were held (see Figure 2 below). The meeting was conducted in India (New Delhi, Guwahati, and Itanagar), Bangladesh (Dhaka), and Singapore, where Bangladeshi, Bhutanese, Chinese, and Indian participants from Track 3 to Track 1.5 were present.

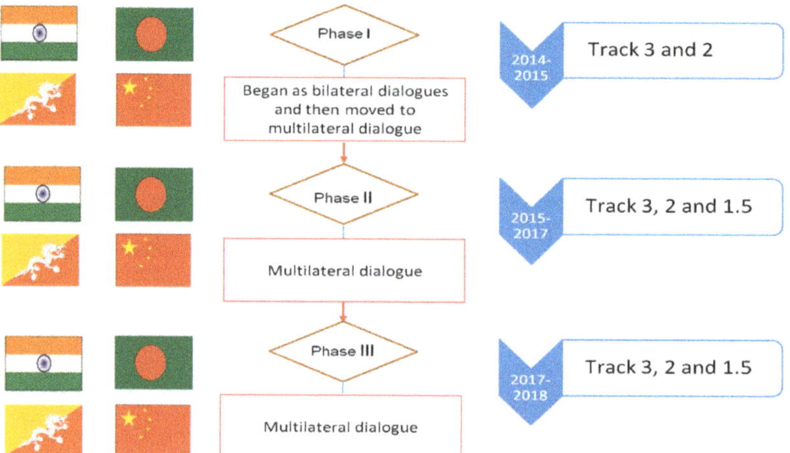

Figure 2. Brahmaputra Dialogue under the South Asian Water Initiative (SAWI).

Track 3 and 2 involved members from CSOs, NGOs like Aaranayak, Centre for North East Studies and Policy Research (C-NES) from India; Bhutan Water Partnership and the Royal Society for Protection of Nature (RSPN) from Bhutan; and Jagrata Juba Shangha (JJS) from Bangladesh. It also involved academic and research institutions like the Institute of Chinese Studies (ICS), Dibrugarh University and Indian Institute of Technology Guwahati from India; BRAC University, Bangladesh University of Engineering and Technology, and the Institute of Water Modelling from Bangladesh; and Shanghai Institute for International Studies (SIIS) and Yunnan University from China. Track 1.5 included government officials from various ministries and departments of India, Bhutan, and Bangladesh. For instance, the Ministry of Jal Shakti (previously Ministry of Water Resources), Brahmaputra Board, Water Resources Department of Assam, and Arunachal Pradesh, Assam Disaster Management Authority from India; Ministry of Water Resources, Bangladesh Water Development Board, Water Resources Planning Organization from Bangladesh; National Environment Commission Ministry of Agriculture and Forest and Ministry of Home and Cultural Affairs from Bhutan. Track 1.5 actors have always been reluctant about their participation in these dialogues, and have particularly stated that the opinions shared on the platform are personal and do not reflect the opinions of the state. There has not been any particular formal statement from a government body, which itself shows a lack of commitment from governance institutions. Track 2 has always emphasized data sharing that can contribute to research activities, and the grassroots and civil groups have emphasized the need for transparency in the decision-making processes of the countries. The Track 3 stakeholders have always emphasized on the need for inclusive governance, with accountability for the issues raised by the stakeholders who rely on the river directly.

We numbered these meeting reports with unique codes and used them in our analysis section. For instance, the first meeting of Phase I of BD is coded as BD(I), 2015 (1). For at least 12 meetings, we also collected the audio recording and notes made during the meetings. The notes and audio recordings helped to triangulate the data of the BD meeting reports.

3.3. Data Analysis

An iterative process of document analysis was used to analyze the BD meeting reports. To make sense of how cooperation between stakeholders emerged in different tracks, we focused on the question:

"What is the data (text of the meeting/workshop documents) telling us about the cooperation?" We read each meeting document in detail, marking keywords, phrases, and sections, and identified how stakeholders wanted to further collaborate or showed signs of cooperation, for instance, how Indian media stakeholders understood the issues of Brahmaputra Basin and how they wanted to further collaborate with media stakeholders in Bhutan and Bangladesh. We analyzed the text in the BD reports where media stakeholders discussed the ways of collaborating with other countries' stakeholders. Similarly, the outcomes of each meeting were also analyzed to understand the cooperation between different stakeholders.

3.4. *Limitations*

In this article, we analyzed only the meeting documents and notes, and there are certain limitations attached to this methodology. Two main limitations are attached to document analysis. First, the documents were produced for a particular purpose and not aligned to a particular research question. The documents did not provide sufficient detail to answer a particular question (in this case to redefine cooperation). However, we analyzed the documents through the lens of cooperation at different tracks, elaborating how the actors have understood cooperation and how each riparian country progressed in terms of cooperation at different tracks between 2014 and 2018. Second, policy documents report on an event in a specific time period; the data did not have the flexibility to present details before and after the event. The documents used in this article only present the views of stakeholders in those particular meetings, and did not highlight how stakeholders behaved after the meetings.

4. Analysis

4.1. *Re-Interpreting Transboundary Water Cooperation Through Brahmaputra Dialogue*

The Brahmaputra Dialogue was initiated as a bilateral initiative with people-to-people diplomacy in 2013, but from 2014, the dialogue shaped into a multi-track and multi-stakeholder deliberation engaging the four basin nations and identifying avenues of cooperation. While dialogue at the Track 1 level, with a top down approach, has always been considered an acceptable form of formal cooperation, this initiative attempted to acknowledge the inclusivity that dialogues at the Track 3 and 2 levels can bring into the decision-making process, as the perspective on the issues plaguing the basin can flow from the bottom to top only when there is accountability to those whose lives and livelihoods are impacted directly by the river. A narrow definition of water cooperation, limited to the Track 1 governmental domain, not only undermines the fruition of the cooperation in other forms, but also actively prevents the maximized impacts being generated. For example, the Indus Treaty has been a diplomatic initiative purely at the political and policy level, but it still remains disputable and unsatisfactory to the basin-level stakeholders on the ground [53,54]. This platform has served to provide a non-formal cooperative arrangement to not only the policy makers and bureaucrats (former and serving) but also to those engaged with civil society and research. The ability to bring on board the serving bureaucrats has been vital, as under certain circumstances formal communication is not possible, but unofficial bodies can facilitate the deliberation among the parties. The initiative has fostered relationships that have lasted beyond the dialogue meetings, and have initiated joint efforts beyond the platform to work together on relevant issues among the stakeholders.

In the first phase, the dialogue moved to a Track 2 mode. The structure of the workshops saw country-level workshops followed by regional-level dialogues. The country-level workshops were conducted only in India and Bangladesh, with plans to expand them to China and Bhutan in the next phase. In the second phase, the dialogue expanded its reach to Bhutan and China by organizing country consultation meetings in both countries, along with dialogue workshops in India and Bangladesh. The third phase has been concentrated on particular themes that were the outcomes of the first two phases—institutional mapping, disaster management, inland water navigation, and water–energy nexus. It has provided space for the government stakeholders to formally and informally deliberate on issues concerning the basin, and received significant participation from China to advocate south–south cooperation on developing water–energy nexus in the region.

The dialogue meetings and workshops conducted have been cross-cut across the tracks in the aforementioned phases, but the outcomes can be outlined at different track levels. Dialogue workshops conducted between 2014 and 2018 are listed in the table in Appendix A. The workshops, meetings, and reports have been provided with a unique code, which were used for reference in the analysis section.

Figure 3 represents the recurring themes that were identified from document analysis and that have been emphasized by different stakeholders. These themes are also cross-cutting and interconnecting. Each theme also identifies which group of stakeholders is more invested in working towards cooperation in the basin through color codes. In the above diagram, "active" denotes energetic pursuit of an activity by being on the forefront, while "passive" involves watching, looking at, or listening to things rather than being actively involved in an activity. This schematic representation is intended to address how dialogue/diplomacy at the informal level can also contribute to cooperation through collaborations at that level and by keeping the diplomats at the formal level diplomacy informed. The details of this representation are addressed in the following.

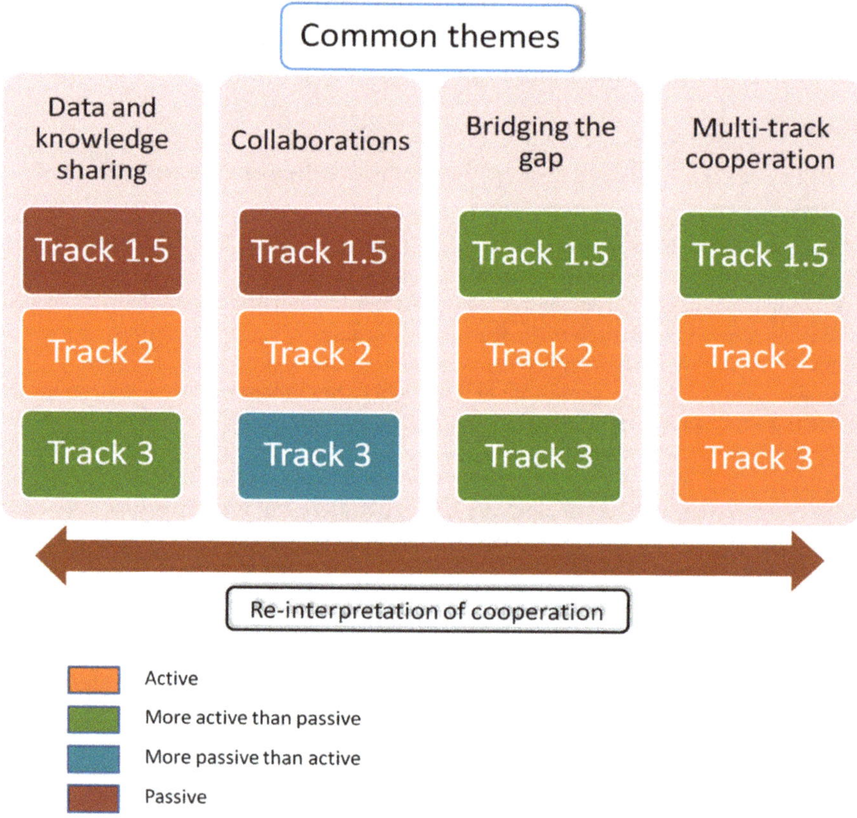

Figure 3. Schematic representation of the analysis.

4.1.1. Data and Knowledge Sharing

Since its inception, the dialogue has involved active participation from the grassroots level (Track 3), as any sort of policy dialogue has to acknowledge the association of the communities that are primarily dependent on the river. "Any intervention implemented in Arunachal should take into account the land and water rights of the tribal communities" [55]. In fact, in 2016, an initiative was exclusively taken under the dialogue to bring together the CSOs of the three countries—the Centre for North East Studies and Policy Research (C-NES) from India, the Royal Society for the Protection of Nature (RSPN) from Bhutan, and Jagrata Juba Shangha (JJS) from Bangladesh. The aim was to bring about cooperation by the sharing of knowledge and experiences between these groups. Within this initiative, a gendered narrative (from marginalized groups) on coping and adapting to disasters across India, Bhutan, and Bangladesh at the community level was also developed, with the intention of sharing the learning with academics/researchers (Track 2) and policy makers (Track 1). "Free flow of data is required, which can contribute to reduce misunderstandings to a great extent" [55].

The best practices followed in Bangladesh, such as community-level disaster management systems, have been shared [56], so the other basin countries can see if these can be incorporated in their nations as well. Bangladesh's capacity to cope with disasters, with efficient communication from top to bottom, was well appreciated by not only the CSOs, but also the government officials that were present from India and Bhutan. The workshop in Shanghai, organized in September 2019, brought together both academics and ex-bureaucrats to discuss how to realize multiple benefits, including optimized energy security and enhanced climate change resilience, through international water cooperation [57]. There are several joint collaborations, mentioned in the following sections, that have also facilitated the sharing of information across. While Track 1.5 has been involved in the process of sharing information, it is usually Tracks 2 and 3 that have fostered better research outputs in the basin. These tracks are also responsible for communicating the information requirements to diplomats from Track 1, who facilitate such transnational data exchange.

4.1.2. Collaborations

A Facebook group titled "The Brahmaputra Dialogue" was initiated in 2017, which brought together the members of CSOs and academics of the three countries (India, Bhutan, and Bangladesh), along with media representatives. This form of cooperation can promote the sharing of information, generating common understanding on various issues related to the river and also building consensus regarding contested issues. Such social media groups can also help in facilitating advocacy at inter-country level [58]. The dialogue has also paved the way for science and media communication initiatives for the basin, engaging scientists and media personnel for improved generation of information and to avoid misinformation, which has been an issue in the region.

The dialogue participants have often emphasized the importance of conducting joint research and how it can help to promote cooperation in the basin. "Joint research should be conducted at the basin level by bringing all the riparian countries together, regarding issues related to the river basin" [59]. Yunnan University has extensive experience working on Mekong River, and they agreed to share their tools, which can also be applied to the Brahmaputra Basin. As a result, a basin-level project was initiated between Yunnan University from China, IIT Guwahati, and the Institute of Water Modelling (IWM) from Bangladesh. The project, titled "Water Resources Vulnerability and Security Assessment of the Yarlung Tsangpo–Brahmaputra Transboundary River Basin" is funded by the National Natural Science Foundation of China and the International Centre for Integrated Mountain Development (ICIMOD). Further, in 2017, a MoU (Memorandum of Understanding) was signed between Yunnan University from China and Indian Institute of Technology Guwahati (IIT Guwahati) to carry out data sharing, and exchange of faculty and students to ultimately foster joint research in the basin.

In 2019, the initiative was taken to develop a book called "Perspectives on the Yarlung–Tsangpo–Brahmaputra–Jamuna River". The book is one of the first of its kind, as it is being written in a collaborative manner by academics from all four riparian countries. The objective of this book is to introduce the multiple dimensions of this river, including hydrology, cultural, biodiversity, development, and so forth. The efforts for collaboration have happened only at the Track

3 and 2 level, with Track 1.5 actors being involved either as experts or recipients of the project outputs, like policy brief, reports, and research papers that can help influence decision making.

4.1.3. Bridging the Gap

Continuous deliberation with the participants through the dialogue has helped to build trust and bring cooperation among academics from India, China, and Bangladesh. The first step towards this was taken in [60], when the first country-level workshop was organized in Yunnan University in China, which also included participants from India and Bangladesh. While discussions and efforts on transboundary water diplomacy and cooperation have been focused on Track 1 cooperation among governments, academic communities also have a crucial role to play. Communication and collaboration between academics can generate many benefits, such as generating and sharing knowledge on water diplomacy and cooperation, developing the capacity of the next generation of water diplomats, identifying opportunities for conflict prevention and cooperation over transboundary water resources, and developing and improving relevant tools. Moreover, in some countries where transboundary water cooperation might be politically sensitive, academic discourses are important to creating social momentum and bridging the communication and understanding among their respective citizens. Under the Brahmaputra Dialogue, an increasing number of academic players have been brought into the conversation. Take China, for example—the dialogue was initiated with only researchers from the Yunnan University Asian International Rivers Center as participants. After several years of development, its network has grown substantially within China to include more academic institutions as well as governmental think tanks. Similarly, in Bangladesh, apart from IWM, academics from BRAC University are now involved in the process. While mistrust has been the roadblock to citizens from different countries from getting to know each other, suspicion among citizens is counterproductive to advancing the cooperation agenda. "The way with trust, confidence, dialogue, and consultation, a major trans-boundary river like the Mekong has come up with a commission, similarly it is possible for Brahmaputra to be the subject of some kind of consensus among its riparian countries—it may take 10–15 years but it is definitely possible through such dialogues" [59]. This Track 2 level cooperation can help to reveal the unknown, which is the first step for dismantling mistrust and promises the hope of reaching Track 1 cooperation in the governmental domain.

Bureaucrats from Assam and Arunachal Pradesh in India have even endorsed the dialogue and recognized the importance of involving multiple stakeholders [59]. "...involving multiple stakeholders at multiple levels from all the basin countries, which will ultimately lead to wellbeing of the common people" [59]. As a result of these continued deliberations, the participants themselves demanded the continuation of the dialogue in 2015 [61]. This, itself, can be seen as a point of cooperation, with the four riparian countries wanting to discuss the issues and concerns through the informal platform. In April 2017, serving and former bureaucrats (along with members of CSOs) from India and Bhutan visited Bangladesh to better understand the disaster management system in place in Bangladesh [56]. In order to facilitate this exchange of information, discussions were conducted with a few of the union- and district-level Disaster Management Committees of Bangladesh. This form of cooperation helped the exchange of information related to disaster management between the three countries. "Various suggestions have come—holistic and basin-level approaches, integrated water resource management, regional cooperation, etc.—but all these will not succeed without dialogues and consultations between riparian countries" [59]. Track 1.5 level diplomacy has also helped to bridge the gap between government officials and civil society [62]. Track 3 and 2 play more active roles than Track 1.5.

4.1.4. Multi-Track Cooperation

By being multi-stakeholder in nature, the dialogue over the years has provided a platform for deliberations of stakeholders such as serving and retired burcaucrats, NGOs, academics and researchers, and CSOs of all the four riparian countries. Therefore, the dialogue has not only helped in building cooperation among the government officials (serving and retired bureaucrats) of the riparian countries but also between officials and other stakeholders. In a group discussion during [58] at the Track 3 level in India, the participants themselves highlighted the importance of cooperation between them. "CSOs also need to motivate themselves into working as a team, whether with other CSOs or with the research community, as one single CSO might not have the capacity to deal with certain problems alone" [58]. Further, CSOs need to engage with the media to highlight important stories and issues. " ... civil society and NGOs working on the ground should be in regular touch with media through e-mail, Whatsapp, and other social media networks so that they can come into parlance with larger issues" [58]. As compared to other river basins of South Asia, Brahmaputra is relatively under-researched [11]. Due to the lack of available scientific information, academics (Track 2) have been a central part of the Brahmaputra Dialogue. Starting with India and Bangladesh, as the dialogue progressed, academics from China also became a part of the process, generating an atmosphere of cooperation among this group multi-laterally [59].

"Cumulative Environmental Impact Assessment Studies at the transnational level should be taken up as a means of cooperation across riparian countries." [59]. When the dialogue became multi-track, the initiative was also presented to major political leaders, such as the Chief Minister of Arunachal Pradesh and the Secretary of Water Resources Department (WRD) Assam, who appreciated the efforts of the dialogue [63]. The members of the Central Water Commission of India and water resource departments in both Bangladesh and India had agreed to be on the advisory board for the next phase of the dialogue by the end of the first phase, making the dialogue multi-track [63]. During the Bangladesh country workshop [56], members of RSPN and C-NES (along with others) conducted field research along with a multi-track meeting, where the discussions concentrated on the local-level management of disasters in the country. The meeting was organized by JJS under the BD initiative, with participation from government departments of India and Bangladesh (Track 1.5).

Academics and researchers (along with other stakeholders) from all the four countries have also come together through the dialogue on various occasions, such as the regional-level workshop in Singapore in October 2016 [64] and the Brahmaputra River Symposium in New Delhi [65] in September 2017, which has helped in enhancing cooperation among them. "Ecological needs must be taken into account when we talk about development and therefore a multi-dimensional approach is needed" has been a suggestion during [59]. Sometimes, academics or academic outcomes may exert influence over a country's political leaders' decisions. For example, the Chinese have shared their experience from Lancang–Mekong on long-term cooperation during the initial dialogues, and how the same strategies could be adopted towards the formation of the Brahmaputra River Commission [55].

In September 2019, IIT Guwahati and the Shanghai Institute for International Studies, a governmental-affiliated think tank, co-hosted the first multi-lateral workshop in China on "Climate-Water-Energy Nexus and South-South Cooperation" with participants from China, India, and Bangladesh [57]. Governmental officials also participated as observers. The workshop discussions highlighted the paramount importance of academic collaboration in creating consistent and positive discourses. Therefore, it is evident that Track 2 academic cooperation has already gained traction, as well as Track 1.5, in all the riparian countries. Therefore, the Brahmaputra Dialogue has helped the development of cooperation among academics and researchers of all the four riparian countries, often shaping it for the coming phases through suggestions that would sustain the initiative and seek to influence policies. For example, [65] and the following consultation meeting with Chinese delegates [66] brought forth recommendations for capacity building of the existing institutions to manage the river system effectively, integrated investment in the Brahmaputra Basin to mitigate risks and make more productive use of water resources, and enhancing cooperation between the riparian countries and states by promoting inland water navigation, finding nodal partners from each riparian nation, institutional mapping, benefit sharing, media involvement, and disaster risk reduction. These suggestions have been integral to the third phase of the dialogue. Track 1.5 seems more eager to participate in the current phase, reflecting political willingness to cooperate, making them more active than passive now.

From the beginning, conducting joint research on issues of common interest has been emphasized across the tracks to ease the sharing of knowledge across the countries, and has been achieved in the more recent phases of the dialogue. Since each riparian nation has a different perspective on river water management, the dialogue has been able to identify common avenues that could generate cooperation, like flood and erosion management, inland water navigation, and the water–energy nexus. Patience is the key to such dialogue projects, as has been emphasized by the stakeholders time and again, to generate willingness to cooperate on a regional level. The Mekong River Commission, which belongs to a more familiar geographical context as both basins are in the South of Asia with familiar development issues, took 37 years to materialize [67]. The dialogue acknowledges the contribution of diplomacy and cooperation efforts at the Track 1.5, 2, and 3 levels as effective and necessary, because the outcomes keep the Track 1 informed. This provides encouragement for the Track 1 diplomats to also engage in basin-level dialogue formally.

5. Conclusions

While state cooperation in transboundary waters is seen as a logical consequence of interdependencies, such cooperation is driven by several factors, such as national security, historical rivalries, hydrological conditions of the basin, and also, at times, intervention of third parties [32]. It is a drawn-out process and, at times, states may not be motivated enough to cooperate. Hence, there is a necessity to expand the focus of cooperation beyond state actors. The Brahmaputra Dialogue provides a neutral platform for open communication among participants. The dialogue does not necessarily focus on a consensus outcome, but is a multi-lateral platform for informal engagement and consultation to identify avenues for cooperation in the transboundary context. Through multi-stakeholder engagement, the dialogue initiative aims to increase cooperation at multiple levels and decrease conflict within the basin. While transboundary cooperation is mostly looked at as a state-led process resulting from political interaction between the riparian countries, this initiative emphasizes the need to widen the scope of cooperation to incorporate initiatives that are happening outside the formal process. Such transboundary interactions between non-state actors could influence resolutions of the transboundary water issues of the Brahmaputra Basin.

Flood management, erosion control, hydropower, navigation, and ecological integrity etc., are issues of high importance to all countries sharing the basin, but there is a need to better understand the system in order to improve its management for economic development. Although researchers, water practitioners, and managers, among others, have conducted substantial analyses to understand the dynamics and potential of this mighty river, there remain significant knowledge gaps in the system and in sustainable approaches able to make the most productive use of rich water resources while reigning in destructive forces. Due to the securitization of hydrological data, there is secrecy around water knowledge in the basin, and a lack of transparency surrounds the knowledge that is available. All of these issues have also resulted in knowledge gaps, which pose a real challenge to IWRM in the region. By bringing the academic community of all the four countries together, this initiative is providing them a platform to interact and work in cooperation to generate basin-wide knowledge. Such basin-wide knowledge can help to strengthen the evidence base and enhance the shared understanding of the system. Such understanding would foster more strategic and cooperative planning across administrative and sectoral boundaries, as well as in multiple disciplines. This, in addition to strengthening the interface between science and policy, would lead to more informed decision making for improved policy formulation (such as the SDGs) and river basin management.

Several focal points where the countries could cooperate have emerged only because the dialogue could be sustained to provide an opportunity for the stakeholders to identify the common issues. Therefore, the dialogue also goes beyond hydrological data sharing or signing of a basin-level treaty, thus broadening the definition of "cooperation" in the Brahmaputra Basin. The identified focal points of cooperation include the academic exchange of scholars, joint research proposals, organizing joint workshops and conferences, joint publications, civil society meets, media interactions, and science–media dialogues. Such collaboration is already paving the way in the Brahmaputra Basin and can be seen as an entry point of cooperation among the Brahmaputra Basin countries.

Author Contributions: Conceptualization, A.B.; methodology, S.V., A.B.; formal analysis, A.D., V.G., X.L.; investigation, A.D., V.G.; resources, V.G.; writing—original draft preparation, A.B., A.D., S.V., V.G.; writing—review and editing, A.D., V.G., A.B., S.V.; supervision, A.B.; project administration, H.M.Q., A.B.; funding acquisition, H.M.Q.

Funding: This research was funded by the World Bank Group W, grant number 7187448 and the APC was not charged from the authors.

Acknowledgments: We would like to express our gratitude to IIT Guwahati, India and SaciWATERs, South Asian Consortium for Inter-Disciplinary Water Resources Studies, Secunderabad, India, for organizational support during the study. The authors are also thankful to Juna Probha Devi, Research Scholar, IITG for preparing the map for the Brahmaputra Basin (Figure 1) and Taylor Warren Henshaw, Consultant to The World Bank for copy editing the paper.

Conflicts of Interest: No potential conflict of interest was reported by the authors.

Appendix A

Table A1. Workshops, meetings and reports from Phase I to Phase III (2014–2018).

Sl. No.	Workshop/Meetings /Reports	Location	Month/Year	Stakeholders Involved	Unique Code
			Phase I		
1	Country-level meeting	India	January 2015	Government and non-government stakeholders from India	BD(I), 2015 (1)
2	Bilateral meetings with government officials in Assam	India	March 2015	From following departments—Flood and River Erosion Management Authority (FREMA), Brahmaputra board, Department of Water Resources, and Department of Environment and Forest.	BD(I), 2015 (2)
3	Bilateral meeting with government officials in Arunachal Pradesh	India	April 2015	From following departments—Department of Water Resources, Department of Forest and Environment, and the Chief Minister's office	BD(I), 2015 (3)
4	Multi-lateral dialogue meeting	Bangladesh	May 2015	The dialogue moved from bilateral to multi-lateral level with the inclusion of stakeholders (track 2 level) from Bhutan and China	BD(I), 2015 (4)
5	Dissemination meeting	India	August 2015	Government and non-government stakeholders from the four countries	BD(I), 2015 (5)
6	Consolidated report	–	–	–	BD(I), 2015 (6)

Table A1. *Cont.*

Sl. No.	Workshop/Meetings /Reports	Location	Month/Year	Stakeholders Involved	Unique Code
			Phase II		
7	Advisory committee meeting	India	February 2016	A committee with mostly academics was formed to forward the dialogue in the respective countries.	BD(II), 2016 (1)
8	Role of dialogue in transboundary water management (Policy brief)	–	February 2016	–	BD(II), 2016 (2)
9	Country-level meeting	Bangladesh	June 2016	Government and non-government stakeholders including Senior Secretary, Ministry of Water Resources, Bangladesh	BD(II), 2016 (3)
10	Bilateral meeting	Bangladesh	June 2016	Non-government and government stakeholders from MoWR, Joint River Commission (JRC), WARPO, Bangladesh Water Board	BD(II), 2016 (4)
11	Multi-lateral country level consultation meeting	China	July 2016	Meeting organized at Yunnan University between academics to identify joint research themes	BD(II), 2016 (5)
12	Country-level meeting	India	August 2016	Non-government and government stakeholders including Secretary MoWR, India, to discuss ways for cooperation among the states within India	BD(II), 2016 (6)
13	Consultation meeting	Bhutan	September 2016	Non-government and government stakeholders of various departments like National Environment Commission, Ministry of Agriculture and Forest, Ministry of Home and Cultural Affairs	BD(II), 2016 (7)
14	Closed door meeting during International River Symposium	India	September 2016	Government and non-government stakeholders from Bangladesh, Bhutan, and India (under Chatham house rule)	BD(II), 2016 (8)
15	Regional-level dialogue meeting	Singapore	October 2016	Government and non-government representatives of four countries including the Senior Secretary, MoWR, Bangladesh	BD(II), 2016 (9)
16	Country-level workshop on Brahmaputra Knowledge Exchange Programme	India	November 2016	Attended by CSOs, academic community and state officials to bridge the knowledge gap on science, policies, and common perceptions about the Brahmaputra River	BD(II), 2016 (10)

Table A1. *Cont.*

Sl. No.	Workshop/Meetings /Reports	Location	Month/Year	Stakeholders Involved	Unique Code
17	Consolidated report from January–December 2016	–	–	–	BD(II), 2016 (11)
18	Country-level workshop	Bhutan	March 2017	National deliberation between state officials and CSOs on transboundary river governance of Brahmaputra River	BD(II), 2016 (12)
19	Country-level workshop	Bangladesh	April 2017	Government and non-government stakeholders for disaster management for the Brahmaputra Basin	BD(II), 2017 (13)
20	Country-level workshop	India	June 2017	Skill and training workshop for Tracks 3 and 2	BD(II), 2017 (14)
21	Regional symposium "Brahmaputra River Symposium: knowledge beyond boundaries'	India	September 2017	150 delegates including government and non-government stakeholders from within and outside the region	BD (II), 2017 (15)
22	Brainstorming meeting between India, China, and Bangladesh	India	December 2017	The discussions during the academic meeting contributed to the understanding of the outcomes of the existing dialogue process, the gaps and challenges associated with it, and the way forward for the third phase.	BD(II), 2017 (16)
			Phase III		
23	Inception meeting for Phase III	India	May 2018	Government stakeholders from India and Bangladesh	BD(III), 2018 (1)
24	Bangladesh country-level meeting	Bangladesh	August 2018	Government and non-government stakeholders	BD(III), 2018 (2)
25	Climate–water–energy nexus and south–south cooperation	China	September 2018	Government and non-government stakeholders	BD(III), 2018 (3)
26	CSO meet for the Brahmaputra River Basin	India	November 2018	Non-government stakeholders	BD(III), 2018 (4)

References

1. Paisley, R.K. International Watercourses, International Water Law and Central Asia. *Cent. Asian J. Water Res.* **2018**, *4*, 1–26. [CrossRef]
2. Barua, A.; Vij, S. Treaties can be a Non-starter: A Multi-track and Multilateral Dialogue Approach for Brahmaputra Basin. *Water Policy* **2018**, *20*, 1027–1041. [CrossRef]
3. Cascão, A.E. Changing Power Relations in the Nile River Basin: Unilateralism vs. Cooperation? *Water Altern.* **2009**, *2*, 245–268.
4. Zeitoun, M.; Mirumachi, N. Transboundary Water Interaction I: Reconsidering Conflict and Cooperation. *Int. Environ. Agreem. Politics Law Econ.* **2008**, *8*, 297–316. [CrossRef]

5. Wolf, A.T. Shared Waters: Conflict and Cooperation. *Annu. Rev. Environ. Resour.* **2007**. [CrossRef]
6. Zeitoun, M.; Warner, J. Hydro-hegemony—A Framework for Analysis of Trans-boundary Water Conflicts. *Water Policy* **2006**, *8*, 435–460. [CrossRef]
7. Vij, S.; Warner, J.F.; Biesbroek, R.; Groot, A. Non-decisions are also Decisions: Power Interplay between Bangladesh and India over the Brahmaputra River. *Water Int.* **2019**, 1–21. [CrossRef]
8. Warner, J.; Mirumachi, N.; Farnum, R.L.; Grandi, M.; Menga, F.; Zeitoun, M. Transboundary "Hydro-hegemony": 10 years Later. *Wiley Interdiscip. Rev. Water* **2017**, 1–13. [CrossRef]
9. Nan, A.S. Track One-and-a-half Diplomacy: Contributions to Georgia-South Ossetian Peacemaking. In *Paving the Way: Contributions of Interactive Conflict Resolution to Peacemaking*; Fisher, R.J., Ed.; Lexington Books: Washington, DC, USA, 2005; pp. 161–173.
10. Montville, J.V. The Arrow and the Olive Branch: A Case for Track Two Diplomacy. In *The Psychodynamics of International Relations*; Lexington Books: Washington, DC, USA, 1991; pp. 161–175.
11. Barua, A. Water diplomacy as an Approach to Regional Cooperation in South Asia: A Case from the Brahmaputra Basin. *J. Hydrol.* **2018**, *567*, 60–70. [CrossRef]
12. Barua, A.; ter Horst, R.; Sehring, J.; Bréthaut, C.; Salamé, L.; Wolf, A.; Pawletta, B.J.; Manzungu, E.; Nicol, A. Universities' Partnership: The Role of Academic Institutions in Water Cooperation and Diplomacy. *Int. J. Water Resour. Dev.* **2019**. [CrossRef]
13. Wehrenfennig, D. Multi-Track Diplomacy and Human Security. *J. Hum. Secur.* **2008**, *7*, 80–88.
14. Hedelin, B. Criteria for the Assessment of Sustainable Water Management. *Environ. Manag.* **2007**, 151–163. [CrossRef] [PubMed]
15. Jackson, R.; Sørensen, G. *Introduction to International Relations: Theories and Approaches*; Oxford University Press: Oxford, UK, 2015.
16. Mearsheimer, J.J. The False Promise of International Institutions. *Int. Secur.* **1994**, *19*, 5–49. [CrossRef]
17. Keohane, R.O.; Martin, L.L. The Promise of Institutionalist Theory. *Int. Secur.* **1995**, *20*, 39–51. [CrossRef]
18. Axelrod, R.; Keohane, R.O. Achieving Cooperation under Anarchy: Strategies and Institutions. *World Politics* **1985**, *38*, 226–254. [CrossRef]
19. Hopf, T. The Promise of Constructivism in International Relations Theory. *Int. Secur.* **1998**, *23*, 171–200. [CrossRef]
20. Wendt, A. Anarchy is What States Make of it: The Social Construction of Power Politics. *Int. Organ.* **1992**, *46*, 391–425. [CrossRef]
21. Selby, J. Cooperation, Domination and Colonisation: The Israeli-Palestinian Joint Water Committee. *Water Altern.* **2013**, *6*, 1–24.
22. Pradhan, S. Water War Thesis: A Myth or A Reality? *Int. J. Arts Humanit. Soc. Sci.* **2017**, *2*, 12–15.
23. Alam, U.Z. Questioning the Water Wars Rationale: A Case Study of the Indus Waters Treaty. *Geogr. J.* **2002**, *168*, 341–353. [CrossRef]
24. Earle, A.; Jägerskog, A.; Öjendal, J. *Transboundary Water Management: Principles and Practice*; Earthscan: London, UK, 2013.
25. Yoffe, S.; Wolf, A.T.; Giordano, M. Conflict and Cooperation over International Freshwater Resources: Indicators of Basins at Risk. *J. Am. Water Resour. Assoc.* **2003**, *39*, 1109–1126. [CrossRef]
26. Wolf, A.T. Conflict and Cooperation along International Waterways. *Water Policy* **1998**, *1*, 251–265. [CrossRef]
27. Wolf, A.T. Transboundary Waters: Sharing Benefits, Lessons Learned. In Proceedings of the International Conference on Freshwater (Hrsg.): Thematic Background Papers, Bonn, Germany, 3–7 December 2001.
28. Grey, D.; Sadoff, C.W. Sink or Swim? Water Security for Growth and Development. *Water Policy* **2007**, *9*, 545–571. [CrossRef]
29. Phillips, D.; Daoudy, M.; McCaffrey, S.; Öjendal, J.; Turton, A. *Trans-Boundary Water Co-Operation as a Tool for Conflict Prevention and Broader Benefit Sharing*; Ministry for Foreign Affairs: Stockholm, Sweden, 2006.
30. Zeitoun, M.; Allan, J.A. Applying Hegemony and Power Theory to Transboundary Water Analysis. *Water Policy* **2008**, *10*, 3–12. [CrossRef]
31. Warner, J.; Zawahri, N. Hegemony and Asymmetry: Multiple-chessboard Games on Transboundary Rivers. *Int. Environ. Agreem. Politics Law Econ.* **2012**, *12*, 215–229. [CrossRef]
32. Leb, C. *Cooperation in the Law of Transboundary Water Resources*; Cambridge University Press: Cambridge, UK, 2013.

33. Cascão, A.E.; Zeitoun, M. Power, Hegemony and Critical Hydropolitics. In *Transboundary Water Management: Principles and Practice*; Earthscan: London, UK, 2010; pp. 27–42. [CrossRef]
34. Bréthaut, C. River Management and Stakeholders' Participation: The Case of the Rhone River, a Fragmented Institutional Setting: River Management and Stakeholder's Participation. *Environ. Policy Gov.* **2016**, *26*, 292–305. [CrossRef]
35. Dore, J.; Lebel, L.; Molle, F. A Framework for Analysing Transboundary Water Governance Complexes, Illustrated in the Mekong Region. *J. Hydrol.* **2012**, *466*, 23–36. [CrossRef]
36. Suhardiman, D.; Giordano, M.; Molle, F. Scalar Disconnect: The Logic of Transboundary Water Governance in the Mekong. *Soc. Nat. Resour.* **2012**, *25*, 572–586. [CrossRef]
37. Suhardiman, D.; Giordano, M. Process-focused Analysis in Transboundary Water Governance Research. *Int. Environ. Agreem. Politics Law Econ.* **2012**, *12*, 299–308. [CrossRef]
38. Sneddon, C.; Fox, C. Rethinking Transboundary Waters: A Critical Hydropolitics of the Mekong Basin. *Political Geogr.* **2006**, *25*, 181–202. [CrossRef]
39. World Bank. *South Asia Water Initiative: Annual Report (English)*; World Bank: Washington, DC, USA, 2017.
40. Ray, P.A.; Yang, Y.C.E.; Wi, S.; Khalil, A.; Chatikavanij, V.; Brown, C. Room for Improvement: Hydroclimatic Challenges to Poverty-reducing Development of the Brahmaputra River Basin. *Environ. Sci. Policy* **2015**, *54*, 64–80. [CrossRef]
41. Singh, S.K. The Brahmaputra River. In *The Indian Rivers: Scientific and Socio-Economic Aspects*; Singh, D., Ed.; Springer Nature Singapore Private Ltd.: Singapore, 2018; pp. 93–104.
42. Gain, A.K.; Immerzeel, W.W.; Sperna Weiland, F.C.; Bierkens, M.F.P. Impact of climate change on the stream flow of the lower Brahmaputra: Trends in high and low flows based on discharge-weighted ensemble modelling. *Hydrol. Earth Syst. Sci.* **2011**, *15*, 1537–1545. [CrossRef]
43. Turner, A.G.; Annamalai, H. Climate change and the South Asian summer monsoon. *Nat. Clim. Chang.* **2012**, *2*, 587. [CrossRef]
44. Gain, A.K.; Giupponi, C.; Renaud, F.G. Climate change adaptation and vulnerability assessment of water resources systems in developing countries: A generalized framework and a feasibility study in Bangladesh. *Water* **2012**, *4*, 345–366. [CrossRef]
45. Babel, M.S.; Wahid, S.M. *Freshwater Under Threat: Vulnerability Assessment of Freshwater Resources to Environmental Change—Ganges-Brahmaputra-Meghna River Basin Helmand River Basin Indus River Basin*; United Nations Environment Programme (UNEP): Nairobi, Kenya, 2008; Available online: http://www.unep.org/pdf/southasia_report.pdf (accessed on 28 August 2015).
46. Barua, A.; Vij, S.; Zulfiqur Rahman, M. Powering or Sharing Water in the Brahmaputra River Basin. *Int. J. Water Resour. Dev.* **2017**, 1–15. [CrossRef]
47. Biswas, A.K. Cooperation or Conflict in Transboundary Water Management: Case Study of South Asia. *Hydrol. Sci. J.* **2011**, *56*, 662–670. [CrossRef]
48. Bandyopadhyay, J. A Critical Look at the Report of the World Commission on Dams in the Context of the Debate on Large Dams on the Himalayan Rivers. *Int. J. Water Resour. Dev.* **2002**, *18*, 127–145. [CrossRef]
49. Mahapatra, S.K.; Ratha, K.C. Sovereign states and surging water: Brahmaputra River between China and India. *Clim. Chang. Sustain. Dev.* **2015**. [CrossRef]
50. Nishat, A.; Faisal, I.M. An assessment of the institutional mechanisms for water negotiations in the Ganges-Brahmaputra-Meghna system. *Int. Negotiat.* **2000**, *5*, 289–310. [CrossRef]
51. Gain, A.K.; Wada, Y. Assessment of future water scarcity at different spatial and temporal scales of the Brahmaputra River Basin. *Water Resour. Manag.* **2014**, *28*, 999–1012. [CrossRef]
52. Huntjens, P.; Lebel, L.; Furze, B. The Effectiveness of Multi-stakeholder Dialogues on Water: Reflections on Experiences in the Rhine, Mekong, and Ganga-Brahmaputra-Meghna River Basins. *Int. J. Water Gov.* **2017**, *5*, 39–60. [CrossRef]
53. Uprety, K.; Salman, S.M. Legal aspects of sharing and management of transboundary waters in South Asia: Preventing conflicts and promoting cooperation. *Hydrol. Sci. J.* **2011**, *56*, 641–661. [CrossRef]
54. Iyer, R.R. Indus Treaty: A Different View. *Econ. Political Wkly.* **2005**, *40*, 3140–3144.
55. SaciWATERs. Multi-Country Stakeholder Dialogue to Understand Issues of Common Interest for Improved Brahmaputra Basin Management. 2015. Available online: http://saciwaters.org/bd/assets/downloads/Multi-country%20Stakeholder%20Dialogue%20meeting%20proceeding.pdf (accessed on 6 December 2019).

56. SaciWATERs. The Brahmaputra Dialogue. In Proceedings of the Multi Stakeholder Dialogue on Disaster Management for the Brahmaputra Basin, Dhaka, Bangladesh, 26 April 2017.
57. Deka, A.; Liao, X. The Brahmaputra Dialogue. In Proceedings of the Climate-Water-Energy Nexus and South-South Cooperation, Shanghai, China, 2 September 2018.
58. SaciWATERs. The Brahmaputra Dialogue. In Proceedings of the Developing Skill and Knowledge—Training Workshop on the Transboundary Brahmaputra River, Guwahati, India, 13–14 June 2017.
59. SaciWATERs. Bilateral Dialogue on the Management of Brahmaputra River Basin. 2015. Available online: http://saciwaters.org/bd/assets/downloads/Dissemination%20Workshop%20proceeding%20for%20Brahmaputra%20Dialogue.pdf (accessed on 6 December 2019).
60. SaciWATERs. Country Level Workshop: China. 2016. Available online: http://www.saciwaters.org/brahmaputra-dialogue/assets/downloads/Country%20Level%20Workshop%20China.pdf (accessed on 6 December 2019).
61. SaciWATERs. Role of Dialogue in Transboundary Water Management. 2016. Available online: http://www.saciwaters.org/brahmaputra-dialogue/assets/downloads/bd-policy%20brief.pdf (accessed on 6 December 2019).
62. Snodderly, D. *Peace Terms: Glossary of Terms for Conflict Management and Peacebuilding*; Academy for International Conflict Management and Peacebuilding (USIP): Washington, DC, USA, 2011.
63. SaciWATERs. Consolidated Report. 2015. Available online: http://saciwaters.org/bd/assets/downloads/consolidated%20report-brahmaputra%20dialogue%20phase%20II.pdf (accessed on 6 December 2019).
64. SaciWATERs. Regional Level Workshop. 2016. Available online: http://www.saciwaters.org/brahmaputra-dialogue/assets/downloads/Regional%20Level%20Workshop%20Singapore.pdf (accessed on 6 December 2019).
65. World Bank; SaciWATERs; TERI; IIT Guwahati. The Brahmaputra Dialogue. In Proceedings of the Brahmaputra River Symposium: Knowledge Beyond Boundaries, Delhi, India, 25–26 September 2017.
66. Gulati, V. The Brahmaputra Dialogue. In Proceedings of the Brainstorming Meeting between India, China and Bangladesh, Guwahati, India, 9 December 2017.
67. Barua, A.; Gulati, V.; Fanaian, S.; Deka, A.; Bastola, A.; Banerjee, P.; Vij, S.; Rahman, M.Z.; Perera, A.C.S. *The Brahmaputra Dialogue, Annual Report*; SaciWATERs: Hyderabad, India, 2016.

Communication

Planning in Democratizing River Basins: The Case for a Co-Productive Model of Decision Making

Tira Foran [1,*], David J. Penton [1], Tarek Ketelsen [2], Emily J. Barbour [1], Nicola Grigg [1], Maheswor Shrestha [3], Louis Lebel [4], Hemant Ojha [5], Auro Almeida [6] and Neil Lazarow [1]

1 CSIRO Land and Water, GPO Box 1700, Canberra, ACT 2601, Australia; dave.penton@csiro.au (D.J.P.); emily.barbour@csiro.au (E.J.B.); nicky.grigg@csiro.au (N.G.); neil.lazarow@csiro.au (N.L.)
2 AMPERES, 38A Nguyen Thi Dieu, District 3, Ho Chi Minh City 700000, Viet Nam; tarek.ketelsen@gmail.com
3 Water and Energy Commission Secretariat, Singha Durbar, Kathmandu 44600, Nepal; maheswor2037@yahoo.com
4 Unit for Social and Environmental Research, Faculty of Social Sciences, Chiang Mai University, 239 Huaykaew Rd, Suthep, Mueang, Chiang Mai 50200, Thailand; louis@sea-user.org
5 Institute for Studies and Development Worldwide (IFSD), 8/45 Henley Rd, Homebush West, NSW 2140, Australia; hemant.ojha@ifsd.com.au
6 CSIRO Land and Water, Private Bag 12, Hobart, TAS 7001, Australia; auro.almeida@csiro.au
* Correspondence: tira.foran@csiro.au

Received: 28 October 2019; Accepted: 21 November 2019; Published: 25 November 2019

Abstract: We reflect on methodologies to support integrated river basin planning for the Ayeyarwady Basin in Myanmar, and the Kamala Basin in Nepal, to which we contributed from 2017 to 2019. The principles of Integrated Water Resources Management have been promoted across states and regions with markedly different biophysical and political economic conditions. IWRM-based river basin planning is complex, resource intensive, and aspirational. It deserves scrutiny to improve process and outcome legitimacy. We focus on the value of co-production and deliberation in IWRM. Among our findings: (i) multi-stakeholder participation can be complicated by competition between actors for resources and legitimacy; (ii) despite such challenges, multi-stakeholder deliberative approaches can empower actors and can be an effective means for co-producing knowledge; (iii) tensions between (rational choice and co-productive) models of decision complicate participatory deliberative planning. Our experience suggests that a commitment to co-productive decision-making fosters socially legitimate IWRM outcomes.

Keywords: co-production; development assistance; hydrological modelling; irrigation; IWRM; rational choice; stakeholder participation; scenario analysis; water governance

---

1. Introduction

Strategic river basin planning consists of a complex, socially ambitious set of knowledge production practices, involving monitoring and assessment, expert-led analysis, and participatory planning [1,2]. This set of practices involves the production and synthesis of knowledge in multiple domains, the interpretation of key messages by policy actors, and the deployment of such messages in planning processes. Because they are complex, resource-intensive, and generally publicly funded, and because actors use them to justify particular investments or development trajectories, river basin planning practices deserve critical reflection.

In this perspective paper, we reflect on the processes and outcomes of two participatory river basin planning initiatives, to which—as process designers, implementers, and observers—we have contributed. One subset of the co-authors implemented an exploratory planning study for the Ayeyarwady river basin, completed in 2018 as a step towards a river basin master plan [3]. Another subset initiated Nepal's first participatory water resources development strategy for the Kamala, a 2050 km^2 river basin in Nepal, an ongoing initiative as of 2019 [4]. These capability-constrained, post-conflict, democratizing settings offer vital insights into the strengths and limitations of approaches to strategic planning. The paper argues that IWRM-based planning requires co-productive models of planning. This argument is based on the authors' reflection on methodological challenges we navigated when designing and implementing river basin planning projects in Myanmar and Nepal.

The concept of strategic river basin planning has evolved since its emergence in the late 20th C. Until the 1990s, it essentially meant long-term infrastructural development planning, with relatively simple social or environmental analysis. Post-WWII water supply infrastructure was planned according to an engineering-oriented paradigm to meet certain objectives (such as irrigated agriculture and hydropower production) [1]. However, events preceding and during the 1990s revealed the social and environmental limitations of such paradigms to river basin development. Thailand and Nepal, for instance, debated the social acceptability of, and alternatives to, large hydropower dams [5,6]. After experiencing major floods in 1993 and 1995, the Netherlands came to recognise the limits of engineering practice, and the value of integrating spatial planning and water management [7,8]. In 1995, Australia announced a cap on diversions from the Murray Darling Basin, to avoid the ecological collapse of Australia's largest river system. At this time, a less fragmented, more coordinated, and systemic paradigm for water resources planning emerged, as reflected in the 1992 Dublin Statement on Water and Sustainable Development [9]. The Dublin Statement is the foundation for an integrated approach to planning water resources, subsequently promoted by development actors during the 2000s under the name of Integrated Water Resources Management (IWRM).

IWRM aspires to improve three "Es": efficiency, equity, and environmental sustainability [10]. It explicitly promotes the integration of multiple stakeholders, disciplines, and spatio-temporal scales [11]. However, within two decades of its emergence, practices to implement IWRM drew criticism for overly optimistic assumptions about how changes to water planning could deliver the three Es in unequal societies such as South Africa [10], Tanzania [12], and Nepal [13], and transboundary regions such as the Mekong basin. Contemporary IWRM is an ambiguous and diverse set of practices [11,14]. It includes top-down, principle-driven variants, as well as local-level, bottom-up, "expedient" versions [15,16]). The basin planning processes we reflect on in this paper have been influenced by highly aspirational IWRM principles, such as formulating a stakeholder-agreed development plan for the Ayeyarwady basin in Myanmar [17], or the desire among water agency professionals in Nepal to identify optimal development strategies using decision support systems. However, to realize—even partially—such aspirations, a river basin development plan would require meaningful participation and collaboration at, and across, multiple levels of governance [14].

Proponents of IWRM in developing countries face two notable political challenges. The first such challenge is asymmetry of knowledge and power, manifested as the uneven distribution of capability and authority between local and national government. Not only are water and land resources unevenly distributed in river basins, the capability of planners, as well as of affected people, is also concentrated at particular levels and locations (and decisions at one location or level can lead to unwanted consequences elsewhere in the system). For example, until recently, the unitary system of Nepal concentrated planning resources and capability at the centre. Power asymmetry can limit the recognition of local interests, knowledge, and socio-technical water management. It can prioritise national-level water resource development preferences. For example, large hydropower and inter-basin water diversion projects for irrigation have dominated planning conversations in Myanmar and Nepal, respectively [18,19]. The challenge of power asymmetry is exacerbated in contexts of data scarcity and uncertainty about river basins as social–ecological systems, where the state of the basin is influenced by political dynamics and narratives, as much as biophysical processes.

A second political challenge consists of the organizational mode by which IWRM initiatives have been delivered to developing countries. The recurring mode has been the international technical assistance project. Project modalities may constrain local interest and institutionalization, particularly when assistance is narrowly channelled. This risk is heightened in settings such as Myanmar and Nepal, which have restructured their water-related agencies and sought to establish new inter-agency water bodies. Although inter-agency coordination is challenging in any context, it is particularly acute in contexts of radical state restructuring, such as Nepal. Compounding these particular challenges, bureaucratic competition [20] and deficits of trust (between non-state and state actors, between lower- and high-level state actors) may arise.

The above conditions and dynamics have influenced the methods and techniques applied to river basin planning—the focus of this paper. Modern basin planning, which pre-dates IWRM, has favoured a particular set of expertise and stakeholders. For example, the expertise of hydrologists, engineers, lawyers, and national government officials tends to outweigh that of citizens, local officials, livelihood specialists, gender and social inclusion experts, and political economists [21]. The expertise and stakeholders favoured may insufficiently represent the breadth of water- and development-related concerns of people in large or complex river basins [22]. Even the production of disciplinary knowledge—such as a basin-scale surface water model—requires within-disciplinary diversity (e.g., rigorous peer-review) to be credible. The consumption of expert knowledge by non-experts requires accountability and transparency among knowledge producers (e.g., about the implications of uncertainty). When expertise, stakeholder concerns, interests, and relevant socio-technical options are inadequately included, legitimacy is compromised [18,23].

How can IWRM-based river basin planners increase the legitimacy of the strategic planning processes they design and facilitate? We engage with this question by reflecting on how particular methodological commitments, evident in our case studies from Myanmar and Nepal, exert influence on knowledge production and stakeholder participation. By "methodological commitments", we mean recurring preferences that we, as planning practitioners or stakeholders, exhibit towards particular methodologies (understood as conceptual models, and associated study designs and techniques). Such commitments can be explicit or tacit. The commitments discussed in this paper are inferences we have made, based on a review of key primary texts (e.g., terms of reference) and participant-observation.

To support such an interpretation, we use two models of how expertise and knowledge inform policy action. The first model is a rational choice model of decision-making [24,25]. In this model, an authorized decision maker (e.g., a minister, or ministerial council) makes decisions which allocate finite public resources so as to maximize societal utility, based on preferences voiced by citizens [26]. This model of expertise and choice assigns high responsibility to credentialed experts, who advise on the consequences of taking different socio-technical options. The second model is a co-productive model of decision-making. This model attaches relatively greater weight to the knowledge of non-credentialed experts. Through collaborative processes involving diverse actors, it seeks to produce *agreement on shared goals*, and to produce *knowledge relevant to achieving those goals* [27–30]. In the second model, authorities make decisions after recognizing, participating in, and responding to recommendations from co-productive processes (Section 3.2).

The two models of decision-making diverge on what constitutes actionable knowledge and how it should be produced (Section 3.2). Our case studies reveal the tensions that arise when both models co-exist in river basin planning. We describe our attempts to negotiate such tensions, and the consequences for planning of such negotiations. Our motivation is thus to reflect critically on the consequences of making particular methodological commitments, including those to which we contributed, for the purpose of improving IWRM-based planning.

Section 2 expands on the development contexts of Myanmar and Nepal. Section 3 then summarizes the original designs of the two river basin planning initiatives in Myanmar and Nepal, showing how their specific features originate in designs by water resource experts affiliated with international development partners. We also describe how specific commitments led to methodological tensions, which challenged us to revise or augment our original designs. Based on the insights from our two cases, Section 4 discusses implications for realizing IWRM-based river basin planning in developing countries. Section 5 concludes with recommendations.

2. Water Resources Development Contexts: Myanmar and Nepal

2.1. Myanmar

Since 2011, Myanmar's partial and contested democratization [18] has led to a notable increase in technical assistance by international development partners, to multiple sectors. In turn, since 2017, such assistance has yielded an efflorescence of water and water-related studies (e.g., [3,31–33]). Development technical assistance involves the promotion, by partners of national expertise (e.g., Australia and other donors in water resources modelling), of cooperation among partners to focus investment (e.g., an Australia–The Netherlands memorandum of cooperation around water resources assistance in Myanmar), as well as competition among partners promoting IWRM-based planning (e.g., Australia, The Netherlands) and those promoting infrastructural development (e.g., China, Japan, Korea).

In 2015, the World Bank initiated a Decision Support System and Basin Master Plan (herein, "Basin Master Plan") project for the Ayeyarwady river basin, as part of a $100 M credit-financed initiative known as the Ayeyarwady Integrated River Basin Management Project. The 2018–2020 Basin Master Plan project is a major investment in evidence-based planning for Ayeyarwady. It aims to deliver a stakeholder-agreed basin development strategy for the Ayeyarwady, under the auspices of the National Water Resources Committee (NWRC), an inter-agency advisory body formed in 2015. Reaching agreement among a diversity of actors and interests is demanding in any context, let alone in the Ayeyarwady basin of Myanmar, where ethnic armed and quasi-state organizations contend for power and recognition against the Union government.

2.2. Nepal

Nepal's development context includes severe political instability that ended in the early 2000s, followed by a decade of negotiation that led to an agreement to create a federal state with a greater voice for citizens in historically marginalized regions of the country, and a state with more explicit commitments to gender equality and social inclusion. Since the enactment of the 2015 Constitution, Nepal has been undertaking a process of state restructuring, involving the devolution of authority and public revenue to seven new provincial-level governments, and 753 new local government bodies, with local and provincial elections held in 2017. The emergence of provincial agencies involved some transfer of authorities and personnel previously assigned to the central government. At the national level, state restructuring involved negotiations to reorganize and consolidate particular ministries, leading to the emergence of a Ministry of Energy, Water Resources, and Irrigation in 2017.

Since the early 1990s, hydropower development in Nepal has been based on principles of liberalization. Private investment, however, has been constrained by a number of financial and institutional risks [34]. In response, the national government has sought to support hydropower development through greater government involvement—identifying important projects, building national schemes and managing hydropower licensing issues [35]. Water and Energy Commission Secretariat (WECS) started a 2018–2021 study to prepare river basin plans and a hydropower development master plan of all river basins of Nepal, supported by strategic environmental and social assessment of these plans. As with the Ayeyarwady Basin Master Plan project, this is a major investment in strategic planning. It includes hydrological modelling, hydropower optimization studies, and strategic environmental and social assessment. The resultant river basin plans are intended to inform the selection of hydropower, irrigation and water supply infrastructure, as well as natural resource management projects in each basin. As with the Ayeyarwady Basin Master Plan project, this initiative will engage with multiple categories of stakeholder, including representatives of affected communities. In both cases, the process by which stakeholder consultations will lead to stakeholder-agreed outcomes is not explicitly stated [17,36].

3. Methodological Commitments and Consequences

3.1. Formulating Strategy

Our two river basin planning initiatives in Myanmar and Nepal had compatible aims and conceptual methodologies, focussed on the participatory formulation of strategy. Strategy refers to "the art or practice of planning the future direction or outcome of something especially of a long-term or ambitious nature" [37]. To strategize means to formulate courses of action to realize development values. Values are topics which matter (or arguably could matter) to an actor [38] (e.g., improving women's access to water, as proposed in Nepal's draft National Water Resources Policy [Section 3.4.1]). To strategize means to articulate goals, major means-to-goal, actions and responsible parties (Figure 1, "Development Pathways"). Strategizing further involves assessing the strengths and limitations of alternative courses of action to reach a goal (Figure 1, "Development Scenarios"). Such assessment can be done using techniques such as multi-criteria analysis (Kamala) and exploratory scenario analysis (Ayeyarwady). Development scenarios which have been prioritized through such assessment would then receive analysis to identify how they could be implemented (i.e., institutional and political economy analysis) [39–41] (Figure 1). In our case studies, strategies have the status of non-binding texts, which may mobilize further action and investment.

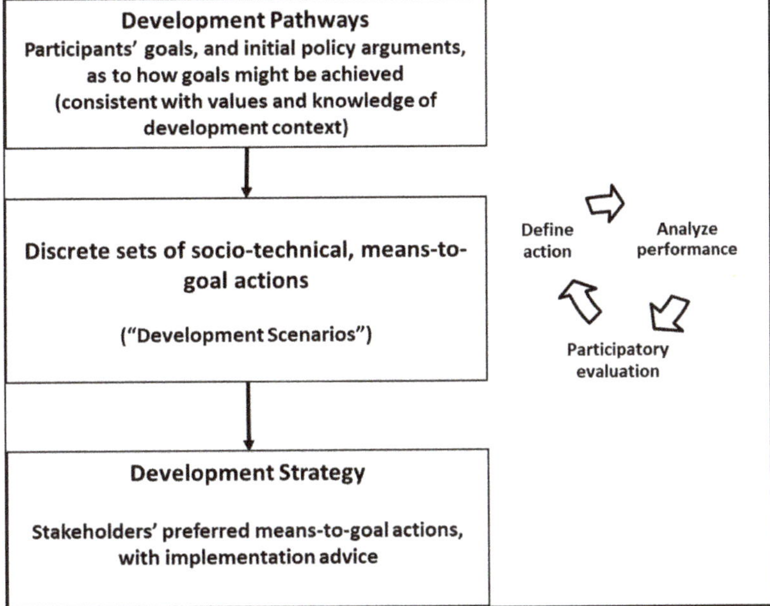

Figure 1. Participatory river basin planning: key components. Source: adapted from [3]. Note: definition of "development pathway" "development scenario", and "development strategy" based on our interpretation of [17]. Note: analysis of performance may use "exploratory scenarios" (Section 3.3).

To support *participatory* formulation of strategy, we anchored both projects to an explicitly deliberative and analytic methodology. By deliberation, we refer to dialogue and argumentation, which aim to generate advice on a set of alternative development strategies or options [38]. An emphasis on deliberation is justifiable, given the weaknesses of participation organized in a top-down, orchestrated manner [42]. Such weaknesses include a tendency to de-politicize values, goals, and means-to-goal actions, for example, by assessing means-to-goal actions using a limited range of evaluation criteria. Accordingly, we reviewed scientific and grey literature on planning approaches which were both technically-informed, and participatory [1,43–46]. In addition, we reviewed literature on specific relevant methods or techniques, such as scenario formulation [3] (chapter 3), multi-criteria analysis [47, 48], and hydrological modelling [3] (chapter 4, 8).

With respect to grey literature, we drew in particular on terms of reference for the Ayeyarwady Basin Master Plan project [17], on the basis that Nepal and Myanmar share broadly comparable development contexts, and common water sector development partners. (The Ayeyarwady Basin Master Plan project also includes the preparation of operational plans, and investment plans—however, these outputs were beyond the scope and resources of the Kamala initiative.)

3.2. Collaborative Model of Governance, Co-Productive Model of Decision-Making

An IWRM-based river basin planning process demands meaningful stakeholder participation. In order to facilitate such participation, we based both initiatives on a collaborative model of governance. Emerson et al. [49] describe collaborative governance as working via three processes, which may interact in a virtuous cycle over time. Each process requires particular kinds of interactions between individuals or small groups:

(i) *Principled engagement.* This refers to interactions that lead to participants understanding each other's interests. Principled engagement requires sufficient initial levels of trust and accountability. It may further emerge through reasoned argument and deliberation, focussed on defining problems and finding agreements together. In order to facilitate reasoned and reflective argumentation, we applied a framework which allows facilitators and participants to distinguish different components of a practical policy argument [3]. Over time, principled engagement enables "shared motivation".

(ii) *Shared motivation.* Shared motivation emerges from interactions that build trust, foster mutual recognition of interdependence, established shared ownership, and create a sense of internal legitimacy.

(iii) *Capacity for joint action.* Joint action refers to mobilisation of knowledge and resources, leading to outputs and outcomes that cannot be accomplished by any policy actor working in isolation, such as recommendations to reform institutional arrangements.

Our river basin planning initiatives sought to catalyse the first two phases of collaborative governance, mentioned above. We proposed designs which were iterative (repeated interactions among a core set of stakeholders); incremental (outputs from earlier activities directly influencing subsequent activities), and deliberative (e.g., use of participatory multi-criteria analysis to support structured argument about water augmentation options in Nepal) (see Figures 2 and 3 below). In doing so, we sought to realize the essence of a *co-productive* model of decision-making in planning, within the limitations of each project (such as language barriers, constrained access to local level stakeholders, and budgetary constraints). In this model, multiple state and non-state actors build knowledge together via processes they value (e.g., processes they regard as credible, legitimate, relevant), leading in turn to outcomes they value (e.g., a strategy regarded as legitimate; citizenship regarded as empowered) [27]. By contrast, in a rational choice model of decision-making, a much narrower group of (elite) policy actors processes information provided by stakeholders and experts, and maximizes societal welfare on the basis of such inputs [24].

Table 1 summarizes the essential components, and important variants, of rational choice and co-production models of decision-making. The models are ideal-types, on a spectrum of models of decision-making. (For example, instrumental versions of co-productive decision-making overlap with variants of rational choice which seek diverse expertise to improve problem and solution framing.) Nonetheless, the models differ with respect to how they conceptualize the process of taking authoritative decisions.

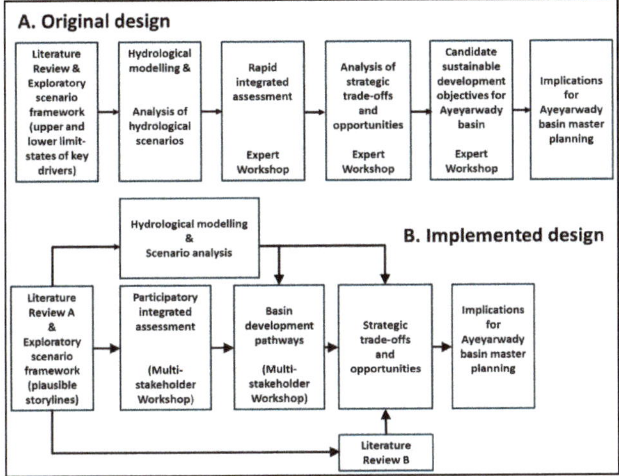

Figure 2. Ayeyarwady Basin Exploratory Scoping Study: original and implemented study designs.

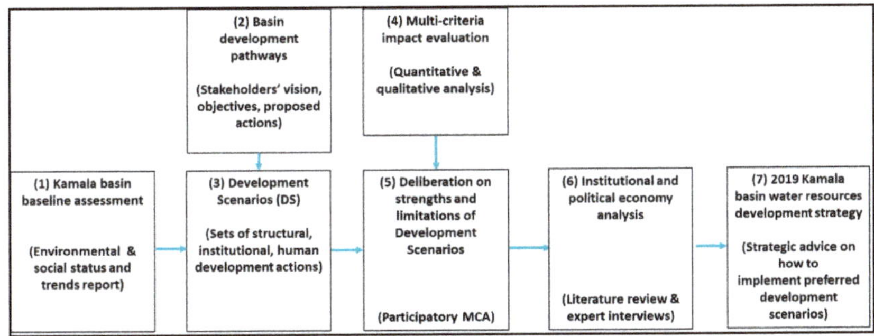

Figure 3. Kamala basin water resources development strategy: implemented study design. Note: Certain proposed actions were not feasible to evaluate using MCA, given resource constraints.

Table 1. Models of decision-making in planning.

Components	Rational Choice Model	Co-Production Model
Typical style of water governance	Hierarchical	Collaborative (network)
Typical governance regime	Centralized	Polycentric
Knowledge-policy linkages	Relatively compartmentalized interactions	Interactions bridging across roles
Assumptions about actors' core roles and capabilities: (A) State authorities (bureaucrats, politicians)	Interpret and process relevant information provided by (B) and (C) Maximize societal utility on basis of inputs from (B) and (C)	Co-design processes Participate in analysis and deliberation Legitimize collaborative processes through recognition, and responding to key outputs
(B) Credentialed experts	Generate substantive findings Translate findings into recommendations	Co-design processes (may lead) Facilitate (documenting outputs & outcomes of key activities) Participate in analysis and deliberation
(C) Non-credentialed experts & lay people	Communicate preferences to (A) and knowledge of impacts to (B)	Co-design processes Participate in analysis and deliberation (may lead)
Time requirements	Low to moderate [1]	Moderate to high [2]
Notable variants	Recognition of, and actions to mitigate, the effect of "non-rational" influences on (A), such as emotions and heuristics	Coproduction of public services Participation as instrumental means

Source: Authors, based on [24,27]. Notes: [1] Proportionate to cognitive complexity; assumes non-conflictual relations between actors. [2] Proportionate to cognitive and political complexity [50].

Table 1 refers to governance regimes [51]. We understand the polycentricity of a governance regime as (i) the degree to which power is distributed among centers of authority (centralized vs. distributed), and (ii) the degree to which effective coordination between authorities exists (highly coordinated vs. fragmented) [52]. A rational choice model of decision-making in planning is typically associated with a centralized and coordinated governance regime, and a co-productive model typically associated with a polycentric (i.e., distributed and coordinated) regime, but such association is not a necessary property of either type of governance regime.

Among the project teams of both planning initiatives (Table 2), the rational choice model of decision-making co-existed with a co-productive model. The following two sections describe the tension between these methodological commitments, and the consequences for river basin planning.

Table 2. Summary: key elements of river basin planning initiatives (this paper).

Element	Initiative 1 (Ayeyarwady Basin, Myanmar)	Initiative 2 (Kamala Basin, Nepal)
Aim	Enhance capability of state and non-state actors to participate in strategic river basin planning by demonstration of relevant techniques	Enhance capability of (primarily) state actors to formulate a river basin development strategy
Objectives	Explore uncertain long-term futures Characterize hydro-physical processes Facilitate dialogue around challenges, goals, and possible development pathways	Define actions and investments to support objectives such as: - improving accessibility, quality and reliability of water resources across the basin; - improving quality of life and environmental standards; - improving representation of women and marginalised people
Timeframe	~12 months (plan) ~20 months (implemented)	~18 months (plan)
Status (October 2019)	Completed	In progress
Focal agency	Hydro-Informatics Centre (HIC)	Water and Energy Commission Secretariat (WECS)
Key participants	Seven national-level agencies Two metropolitan agencies Six non-state organizations 18 Young Water Professionals	Seven federal agencies 12 local agencies One provincial agency Two non-state organizations (as of June 2019)
Focal agency control over implementation	Relatively low	Relatively high
Methodology:		
Model of decision-making in planning	Rational choice (original design) Collaborative (implemented design)	Collaborative (original design) Hybrid[2] (implemented design)
Development pathways	Core focus: Formulated by stakeholders (two rounds of workshops)	
Exploratory scenarios	Core focus: three narrative storylines illustrated with hydrological model runs	Secondary focus (desktop analysis)
Development scenarios	Not in scope	Core focus (four scenarios illustrated with hydrological model runs)
Participatory assessment	Rapid integrated assessment workshop (March 2018)	Multi-criteria analysis workshop (May 2019)
Institutional & political economy analysis	Secondary focus (literature review)	Core focus (literature review & primary data collection)

[2] Refers to co-existence of both models (Section 3.4).

3.3. *Initiative 1: An Exploratory Scoping Study for the Ayeyarwady Basin, Myanmar*

3.3.1. Origins and Actors

The Ayeyarwady Basin Exploratory Scoping Study (BESS) was designed to help bridge the gap between the Ayeyarwady State of the Basin (SOBA) assessment, and the World Bank-supported Basin Master Plan project. BESS was sponsored by the Australian Water Partnership, Myanmar Directorate of Water Resources and Improvement of River Systems (DWIR), and the Australian science agency CSIRO. The focal agency, HIC, is a project-based entity affiliated with the National Water Resources Committee (NWRC). NWRC is an inter-agency group formed in 2015 to advise on water-related risks and development. Implementing partners consisted of Chiang Mai University—Unit for Social and Environmental Research (CMU-USER), eWater Ltd., Flow Matters Pty Ltd., International Centre for Environmental Management (ICEM), and CSIRO.

With respect to participation in BESS, given the project's exploratory scope, advisors to the Basin Master Plan project recommended that a small subset of stakeholders, based primarily in Yangon and Nay Pyi Taw, be recruited. Participation in BESS primarily meant contributing to structured small group discussions in two one-day workshops: a rapid integrated assessment workshop (March 2018), and a development pathways workshop (May 2018). The agreed participant pool consisted of four categories of actors:

(A). Advisory Group (AG) to the National Water Resources Committee. The advisors comprise a small number of senior experts, some of whom are former government officers;

(B). Officers and young professionals affiliated with HIC's Young Water Professionals (YWP) and Junior Researcher programs, whom we invited and trained to serve as rapporteurs and task facilitators for project workshops;

(C). Officers of government agencies other than DWIR. We invited agencies responsible for domains of central relevance to strategic basin planning to nominate technical or policy experts to participate in the project workshops. Those domains included hydropower, forest conservation, irrigation development, pollution control, rural water provision, and urban planning;

(D). Representatives of civil society and research organizations. Criteria for recruitment included organizations with an interest in strategic water-related planning, or with a reputation for previous substantive work on a related topic, or having a prior contribution to the SOBA. We invited both domestic and international organizations, with a preference for Myanmar nationals.

For participation in the final, development pathways workshop, all attendees of the first workshop were invited. In addition, a small set of new organizations and individuals were invited, selected on the basis of their expertise on issues the team considered important (e.g., agricultural development).

3.3.2. Methodological Commitments

In addition to its capacity building objectives, sponsors conceived of BESS as an opportunity to provide an independent, expert perspective on risks, opportunities, trade-offs, and synergies in the Ayeyarwady basin [53]. BESS aimed to demonstrate the complex social-ecological implications of resource development decisions. In so doing, BESS could demonstrate the value of an inclusive, integrated human-environment approach, and offer guidance on the development of the subsequent Basin Master Plan project.

Specifically, designers wanted BESS to offer a consolidated understanding of the basin as a hydro-ecological system. This understanding would include the degree to which hydro-physical changes would lead to a change in ecological functioning. The hydro-physical changes of interest included change to land use/land cover (notably from forest conversion or restoration) and changes to flow regimes from the development of water storage for hydropower and irrigation. The effects of interest included impacts on sediment dynamics, fisheries, flooding, and navigation. This consolidated understanding was of interest to BESS' sponsors, because they anticipated that the Basin Master Plan project would emphasize the extent of unrealized natural resource development opportunities, in a manner similar to the Basin Development Plan program of the Mekong River Commission [54].

As shown in Figure 2 below, the original design of BESS focussed on hydrological scenario analysis as an entry point to an understanding of environmental and social impacts. The hydro-physical scenarios would inform a rapid integrated assessment, based on expert judgement. This assessment of ecological and social responses—to be provided in a semi-qualitative manner by specific experts who had co-authored the SOBA—would then allow the team to draw conclusions about development opportunities and trade-offs for the Basin. In addition, the design called for the study team to formulate "candidate" development objectives, for example, water levels that should be maintained to allow year-round navigation between key cities. The envisioned timeframe for the study was approximately 12 months, with the intent of influencing the design phase of the DSS/Basin Master Plan project.

3.3.3. Consequences for River Basin Planning

The original design and methodology for BESS had an ambitious goal (the synthesis of key development opportunities and trade-offs); a broad scope yet short timeframe; and a particular emphasis on hydro-ecological impact assessment. In attempting to implement this design and methodology we experienced several constraints, tensions and contradictions, which are notable from the perspective of IWRM. These challenges required that we adapt and augment the original methodology, as described below.

The original design emphasized hydrological modelled scenarios and hydro-ecological knowledge. The Ayeyarwady hydrological model allowed us to explore the impacts on surface water resources availability under three exploratory scenarios (described below), and to describe notable alterations to flow regimes at the sub-basin level [3].

To conduct the above analysis, it was necessary to translate qualitative narratives to quantitative modelled scenarios. We made modelled outputs available to participants via an online dashboard, presentations from modellers and access to modellers for questions during workshops. However, we experienced several challenges. Some aspects of particular interest to stakeholders (e.g., navigability) were not explicitly modelled. Model parameterisation also required assumptions that could only be partially tested, given the scarcity of observational data [3] (chapter 4). For example, configuring storages in the model required assumptions about operational rules, yet these details are not always known for existing storages, let alone planned facilities. The impacts of hydropower development on downstream flows were sensitive to such assumptions. We made assumptions that stakeholders and modellers regarded as appropriate, choosing operating rules that maximise the impact of hydropower options on the seasonal hydrograph. Such challenges contributed to our decision to complement the modelling with more intensive stakeholder involvement in the impact assessment process.

Time delays incurred during the activities above meant that we were not able to enlist a sufficient number of the original SOBA co-authors, to conduct the rapid integrated assessment (essentially, an estimate of alternative bundles of ecosystem goods and services that the Ayeyarwady basin could provide under the exploratory hydrological scenarios). This delay had implications for the production of the rapid integrated assessment (Figure 2, "A") and for overall project delivery. In order to deliver, within an acceptable timeframe, an assessment of strategic trade-offs and opportunities—a key intermediate output—it was necessary to make several interlinked revisions to the design. These revisions amounted to a move from a multi-disciplinary to a transdisciplinary practice [30].

First, to conduct a rapid ecological and socio-economic impact assessment, assumptions about future states were required to complement the hydrological modelled scenarios. We therefore developed three "Ayeyarwady 2055" exploratory scenarios. These storylines depicted imagined future conditions in upland and lowland zones of the Ayeyarwady basin, based on an initial set of drivers and outcomes that were explicitly social and political, as well as bio-physical. The storylines conveyed essential social and economic dynamics. Although the hydrological scenarios remained analytically distinct, we conceptualised them as devices to illustrate the storylines [3] (pp. 9–44). In so doing, we made them more accessible to a non-specialist set of participants.

Second, the original design focused heavily on hydrological scenarios and made the expertise of the SOBA authors prerequisites for stakeholder discussion about strategic issues. This was consistent with a rational choice model where authority is vested with experts; however, we revised the technique to be more conducive to co-production of knowledge. Our first workshop invited a set of state and non-state stakeholders in Myanmar to discuss such issues using their existing knowledge (Figure 2).

In so doing, we mobilized a wider set of expertise, beyond that of disciplinary specialists (cf., [45]). To motivate this discussion about impacts, risks and opportunities, participants first read and discussed the Ayeyarwady 2055 scenario storylines. The integrated assessment workshop elicited a wide range of development issues which the participants regarded as significant. Participants provided insights regarding the broader, cross-sectoral impacts ssociated with upland and lowland resource development [3] (pp. 45–51).

Our subsequent "development pathways" workshop (Figure 2, "B") invited a broader cohort of participants than was conceived in the original design to formulate major sequences of public actions to attain development objectives, to which, as citizens of Myanmar, they might reasonably aspire [3]. Participants formulated objectives to improve water quality, upland forest and catchment governance, sediment management, electrification, and agricultural development. They articulated ambitious goals and proposed reasonable and relevant sets of objectives linked to each goal. For each objective, they generated actions, ranging from concrete, incremental steps, achievable in the short term, to more complex actions. Some of the latter are ambiguous, and will require clarification and elaboration. Many of the pathways require transformative changes (for example, peace agreements with inclusive approaches to upland catchment management) [3] (pp. 52–60). The study then explored the gap between such aspirations, and dynamics-as-usual trajectories of development in the Ayeyarwady basin. It concluded by reflecting on the implications of its methodology for the ongoing Basin Master Plan project.

3.4. *Initiative 2: Water Resources Development Strategy for Kamala Basin, Nepal*

3.4.1. Origins and Actors

Government of Nepal's Water and Energy Commission Secretariat (WECS) formulated a National Water Resources Strategy in 2002, and a National Water Plan in 2005. WECS provides other government agencies with a technical review of their development plans. Since the early 2000s, it has been the proponent of IWRM in Nepal, with IWRM officially recognised in the Water Resources Strategy [55].

In 2017, WECS intiated the formulation of a National Water Resources Policy. The draft policy proposes differentiated responsibilities for water management, consistent with the 2015 Constitution. Among other things, it proposes that a revitalized Water and Energy Commission and its Secretariat approve the periodic submission of strategic river basin plans. Further, it proposes that all three tiers of government, including the private sector, need to take techno-economic clearance (consent) from WECS to implement any new water resources development project in a river basin. WECS consequently identified that it would need to build hydrological and associated modelling capabilities, in order to support a techno-economic review of proposed water resource development plans and projects. For improving the capacity for multi-objective optimization for planning, as well as participatory planning, the Kamala river basin (population 610,000) was selected as an appropriate case. The implementing partners are WECS and CSIRO.

The participant pool for Kamala basin strategy formulation initially consisted of the following categories of actors, with A–F identified through a stakeholder analysis:

(A). Representatives of each of the Basin's 15 local government bodies (elected officials and/or senior staff);
(B). Representatives of approximately six national government agencies serving on the Advisory Committee to the project;
(C). Academic and consultant water and agricultural professionals;
(D). Officers of the Kamala Irrigation Project, Department of Water Resources and Irrigation;
(E). Civil society organizations representing ultimate beneficiaries;
(F). Representatives of private enterprises operating in the basin, and;
(G). Individual women and men, representing particular communities of ultimate beneficiaries.

As of 2019, actual participants have come from categories A–D. These participants contributed to structured small group discussions in one or more of the following workshops: six visioning and goal setting workshops (July and November 2018), and a multi-criteria analysis workshop (May 2019). By contrast, participation of actors from categories E–G has occurred through interviews and focus group discussions on selected topics (e.g., livelihoods and water use), as opposed to direct participation in strategy formulation.

3.4.2. Methodological Commitments

Figure 3 shows key steps in the methodology as of 2019. Those include participatory formulation of development pathways (following techniques used in Initiative 1); multi-disciplinary analysis impact assessment of development scenarios; participatory multi-criteria analysis (MCA) of development scenarios; and institutional and political economy analysis of preferred development scenarios, for the purpose of providing implementation advice. (The original study design [4] included additional techniques which the study team elected not to pursue because of resource constraints.)

The implemented design is consistent with a deliberative, analytic approach to strategy formulation. However, its detail and complexity has not been well suited to an operating context constrained by time, geographic distance, and disciplinary backgrounds. Furthermore, the scope of topics of interest to participants has been broad, in comparison with the project's resources and timeframe.

3.4.3. Consequences for River Basin Planning

Project constraints drew to light methodological tensions within the project team, as well as between stakeholders, where time, budget, resources and significant uncertainty dominated. As we explore in this section, tensions over prioritization included the acquisition of new data; the production of knowledge by technical experts; and the integration of stakeholder knowledge and participation in the planning process.

The complexities of design and technique led to disagreement across the project team about the significance of each technical component, its definition and necessity, and, hence, its relative priority in a context of constrained resources. For example, one position observed among the team was the prioritization of detailed baseline assessment, as well as the projection of future water demand and supply (Figure 3, step 1), as an important pre-requisite to strategy formulation. This position questioned the value of developing a strategic plan based on limited observational data. A contrasting position observed among the team was a belief that the central methodological challenge was, notwithstanding the inevitability of limited data, could any strategic advice be offered regarding two or more options to achieve key objectives (e.g., meeting agricultural water demand)?

One element in our response to this planning challenge was to use participatory multi-criteria analysis (MCA), a technique that could be iterated to incorporate new knowledge (Figure 3, step 5). Participants in the Kamala initiative had previously formulated three goals as part of their Development Pathways (Figure 3, step 2). The Pathways were formulated through the exchange between local and national government actors of values, goals, and means–goal actions. One of the three goals was a broad, water-centred development goal: "reduced impact of water induced disasters, and improved availability, use, and allocation of water resources for livelihood generation, well-being and economic growth" [56]. Participants envisaged several major actions that could meet this goal: building small or medium reservoirs; the conjunctive use of groundwater and surface water; and building a large inter-basin water transfer scheme (the Sun Koshi–Kamala multi-purpose diversion project [57]). The project team elaborated these actions, plus the rehabilitation of the existing Kamala Irrigation Project [58], into four Development Scenarios. Notwithstanding uncertainty about the status and trends of water supply and demand, the project partners eventually agreed to proceed with a participatory MCA workshop focussed on the Development Scenarios.

MCA is sensitive to prior understandings of a policy issue, and how particular options are described. The question of who evaluates has consequences for representation and legitimacy [48,59]—as such, we recommended inclusive participation. Our MCA design, adapted from [47], involved an impact evaluation conducted by the project team (based on desktop analysis, hydrological modelling, and expert interviews), for nine initial evaluation criteria. Participants individually weighed each criterion twice, before and after viewing evaluation results. The team computed individual utility scores. The median and distribution of individual scores for each Development Scenario were then viewed, as a contribution to the deliberation over prioritizing the Development Scenarios.

The focal agency regarded particular NGOs and research organizations as inappropriately politicized, or insufficiently prepared to engage in such deliberation. Consequently, we invited participants from the three levels of government, and a restricted set of research organizations. Such outcomes reflect unfamiliarity with public participation in the direct formulation of strategy (Figure 3). They further reflect relatively low trust, and perhaps relatively low mutual accountability, between state and non-state actors in Nepal's water sector (cf. Section 3.2). Such dispositions were concentrated among Nepali experts advising the focal agency, but at times were expressed by elected representatives. (In late 2018, some elected local government representatives told us that time constraints prohibited their direct participation in co-production of the water resources development strategy. Instead, they requested to review the team's analyses and recommendations, consistent with a rational choice model of decision-making. The local representatives did not express preferences regarding the participation of other actors)1. We reflect on the appropriateness of the methodological design in the following section.

4. Discussion

4.1. Negotiating Mixed Methodological Commitments

Figure 4 summarizes the consequences for river basin planning of the methodological commitments made in our Nepal and Myanmar initiatives. It summarizes the key influences on those commitments, as well as some responses to mitigate certain undesired consequences.

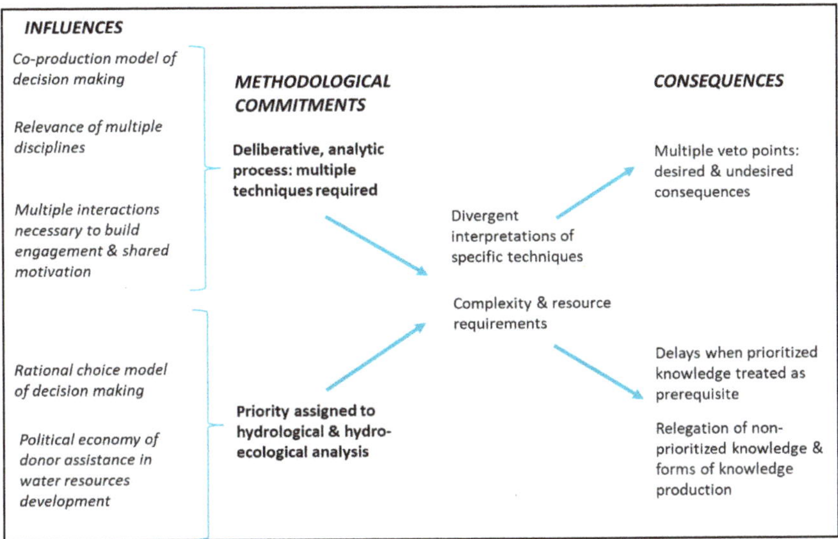

Figure 4. Methodological challenges for river basin planning in developing countries.

In both cases, more than one model of decision-making influenced original design commitments. In both cases, following a co-production model, one subset of the partners committed to using multiple disciplinary techniques to support an analytic deliberative process. On the other hand, consistent with a rational choice model, a second subset of the partners regarded specific expertise to be a prerequisite for participatory basin planning to commence. In both cases, the ensuing epistemological tension led the partners to occasionally disagree over technique prioritization and resource deployment. Both cases highlight the importance of fostering competencies and capacities to negotiate acceptable outcomes when such tensions arise.

In the Ayeyarwady case, challenges in producing and interpreting hydrological modelled scenarios offered the team an opportunity to adapt the rapid integrated assessment method (Figure 2). We widened it from a realtively narrow Delphi process to a process more inclusive of our workshop participants' social, political, and environmental knowledge. This adaptation met the project's overall objectives, but in a manner consistent with a co-productive model, not a rational choice model, of decision-making. By contrast, in the Kamala case, divergence over who could effectively contribute to MCA (Section 3.4.3) impacted discussion around the *pros* and *cons* of the four Development Scenarios.

When mixed methodological commitments lead to undesired consequences from an IWRM perspective, the specific autonomy of process designers matters. In the Myanmar case, our autonomy as process designers was relatively high, whereas in the Nepal case, it was relatively low (as manifested in constraints on the type of stakeholder, and types of knowledge production, admitted or preferred by the focal agency; Section 3.4.3). We interpret this difference in control primarily to contextual differences between Myanmar and Nepal. The Ayeyarwady BESS initiative-an exploratory study with lower stakes for the focal agency–allowed innovation with co-productive methods, with minimal contestation from advisors to the focal agency.

By contrast, the Kamala initiative aimed to produce an actual strategy, and to do so in a context of relatively profound state restructuring compared to Myanmar. Nepal's federal structure has constituted and mobilized local and provincial government actors. Concurrently, key organizations of federal government are in the process of restructuring, along with legislation reform. WECS is proposed to have significant institutional reforms in order to implement IWRM-based river basin planning. It takes time to adopt a co-production model of decision-making and a vigorous multi-stakeholder process. A deliberative analytic design (Figure 3) offers multiple points for project sponsors to intervene, occasionally exercising veto power, to shape the scope and extent of stakeholder participation (Figure 4).

To navigate undesired consequences when multiple techniques are required (Figure 4), we found it helpful to simplify essential techniques (Figure 3), and to propose planning processes that do not allow any one actor to dominate methodological choices (Ayeyarwady). In the Ayeyarwady case, the explicitly co-productive approach mitigated the risk of over-weighting biophysical analysis. In both cases, we found it helpful to affirm and elicit expertise from different disciplines. The ability to iterate contested or unimplemented techniques would further contribute to a shared understandinging of their value.

4.2. Implications for Realizing Aspirations of IWRM

We noted at the outset that IWRM-based river basin planning is a project with high aspirations, such as formulating a stakeholder-agreed development plan for the Ayeyarwady basin in Myanmar [17], or producing a river basin development strategy for the Kamala to realize an optimal set of economic, social, and environmental benefits [60].

Our attempts to further such aspirations led us to adopt analytic, deliberative methodologies, consistent with collaborative and co-productive models of decision-making. However, these are not without constraints. For example, authority and influence in water policy are concentrated among actors who are familiar with stakeholder consultation in the form of roundtable or town hall style meetings, but less familiar with the notion of empowering stakeholders to formulate strategy-based on analytic, deliberative methods. Furthermore, the institutions of river basin planning are nascent in developing, post-conflict settings.

A more subtle constraint arises from the fact that IWRM-based planning is logically consistent with a co-productive model of decision-making (and a collaborative style of governance)—yet, at a global level, development partners promoting IWRM-based planning predominately subscribe to a rational choice model of decision-making (and a hierarchical style of governance). They recognize the instrumental value of stakeholder consultation, but ultimately assign to specialists the task of formulating the options presented to decision takers (see e.g., design documents for the Myanmar BESS project [53] and the DSS/Basin Master Plan project [17]).

This presents a tension for development donors. On the one hand, a state-centric and hierarchical interpretation of IWRM is expedient, and may appear to be the only overarching model of technical assistance acceptable to recipient states. One the other hand, we have shown that contrasting understandings of how expertise links to action, and of what expertise should be prioritized, have important socio-political effects, some of which can undermine the inclusive aspirations of IWRM.

5. Conclusions: Advancing Collaborative, Co-Productive IWRM

This paper focussed on IWRM-based planning in Myanmar and Nepal, during a period in which we had the privilege of co-designing and implementing river basin planning initiatives (2016–2019).

During this period, the politics and institutions related to water resources development were dynamic and contested in both countries. We did not attempt a comprehensive diagnosis of the institutional and political constraints to realizing IWRM aspirations. Instead, by focusing on planning methodology *in action*, we offer insights to improve river basin planning practice in democratizing settings.

Both our planning initiatives made use of participatory, deliberative, analytical designs. Although such designs are consistent with a co-productive model of decision-making, they are complex. In a context of hierarchical governance, such designs offer many intervention or veto points. From the standpoint of realizing the aspirations of IWRM in democratizing settings, such points of interception have desirable and undesirable implications (Section 4.1). Nonetheless, when they apply co-productive knowledge production in and beyond the project cycle, donors and designers advance IWRM-based river basin planning.

Sponsors uncertain of participatory deliberative processes might draw confidence from the relative success of the multi-stakeholder visioning ("development pathways") components in our planning initiatives. In both cases, participants articulated broad development agendas for their river basins [3,56]. They expressed concern for marginalized people (e.g., landless farmers; people in conflict zones), and recommended equitable and ecologically sensitive investment in upland natural resources and farming. They proposed multiple water resources development options (and, in the Kamala, Nepal case, subsequently deliberated on them in detail). They believed their understanding of strategic issues improved, and that co-participants gave reasoned arguments for positions. Participants considered it important to contribute to river basin planning beyond their specific expertise—that is, beyond their instrumental value. They explicitly valued a collaborative multi-stakeholder approach to planning. Participants' evaluations from the Kamala development pathways and MCA workshops are consistent with evaluations obtained by the BESS project [3] (Annex 3).

Donors and designers can scrutinize IWRM technical assistance for the presence of rational choice models of decision-making, in order to reflect on advantages and limitations of their specific models. Sophisticated rational choice models recognize the potential for non-rational influences on decision-maker cognition. They recognize that it is not enough to present credible analysis to public actors who are assumed capable of acting on such knowledge to further their interests, in accordance with their values. However, even sophisticated rational choice models may underestimate the effect of power imbalances between actors, and externalize the challenges of political accountability (e.g., by assuming that elections or other accountability processes will ultimately steer authorities to maximize societal utility).

In the wider Mekong region, much authority to take decisions remains with agencies with the expertise to control floods, irrigate arid land, and otherwise meet water demand through infrastructural means [21]. However, consistent with IWRM principles, water resources development has grown considerably more complex in recent decades. It is evolving from technical optimization to integrated, systemic, foresight-oriented concerns [1]. Yet, state water agencies in the Mekong continue to privilege the original set of engineering and hydraulic expertise.

A co-productive model of knowledge production widens the scope of relevant expertise. The collaborative mode of governance underpinning it explicitly seeks to empower and mobilize civil society and private sector actors. We found that actors thus empowered can set the planning agenda. In the Ayeyarwady case, we found that co-production allows stakeholders to mitigate certain challenges encountered in scenario modelling (Section 3.3.3), and helps shape a planning process that can start immediately, even as data collection and model development continue in parallel. By working with planning stakeholders, we discovered more effective ways to combine and sequence disciplinary knowledge (Figure 2).

Our experiences in Myanmar and Nepal show that, with adequate support, stakeholders can co-design more vigorously, for example providing guidance for which disciplines are necessary and sufficient to address their agenda. We have shown the difficulties of trying to implement co-productive and collaborative models. Yet the alternative—managing complexity and contestation via bureaucratic modes of governance—offers no greater likelihood of planning outcomes legitimized by society (cf., [18]). Development partners instead can point to longstanding practices of collaborative planning in their own contexts. Australia, for example, can offer diverse models of collaborative governance, as reflected in the work of Landcare, Catchment Management Authorities, regional natural resource management bodies, and participatory urban planning initiatives. Such domains have yielded productive multi-stakeholder deliberative initiatives [47,61,62]; comparative insights on collaborative management [63,64]; insights on the challenges of integrating Indigenous knowledge into river basin planning [65]; and reflections on the interaction between collaborative and hierarchical governance [66,67].

In conclusion, methodologies are not neutral in their effects—they empower some actors at the expense of others. IWRM initiatives in democratizing developing countries face multiple challenges. While the rational choice model of decision-making which persists in IWRM-based planning offers an administratively simpler approach to development assistance, it runs the risk of unduly concentrating expertise and power. In so doing, it undermines the ultimate aspirations of IWRM, which require co-productive approaches. We hope the insights offered as a result of our experience can guide improved IWRM investments and outcomes.

Author Contributions: Conceptualization, T.F.; methodology, T.F.; writing—original draft preparation, T.F.; writing—review and editing, T.F., D.J.P., M.S. (Nepal), T.K., E.J.B., N.G., L.L., H.O., A.A., N.L.; project administration, T.F., M.S. (Nepal), D.J.P., A.A., N.L.

Funding: This perspective piece contributes to the South Asia Sustainable Development Investment Portfolio supported by the Australian Aid progam. It was co-funded by CSIRO Land and Water.

Acknowledgments: We thank three anonymous reviewers, Saumitra Neupane, William Young, S.M. Wahid, and Susan Cuddy, for comments provided during the writing process. We acknowledge the Advisory committee for Kamala Basin Initiative, the Australian Water Partnership, eWAter Ltd., Jalsrot Vikas Sanstha, and Policy Entrepreneurs Inc.

Conflicts of Interest: The authors declare no conflict of interest. The funders had no role in the design of this communication piece; in the collection, analyses, or interpretation of data; in the writing of the manuscript, or in the decision to publish the results. Research carried out in the Kamala basin is a collaborative work between WECS and CSIRO.

References

1. Guy, P.; Li, Y.; Tom Le, Q.; Robert, S.; Li, J.; Shen, F. *River Basin Planning: Principles, Procedures and Approaches for Strategic Basin Planning*; UNESCO: Paris, Frence, 2013.
2. Cook, C.; Bakker, K. Water security: Debating an emerging paradigm. *Glob. Environ. Chang.* **2012**, 22, 94–102. [CrossRef]
3. Foran, T.; Grigg, N.; Barbour, E.; Wahid, S.; Gamboa Rocha, A.; Hunter, R.; Sawdon, J.; Moolman, J.; Adams, G.; Rahman, J.; et al. *At the Heart of Myanmar: Exploring Futures of the Ayeyarwady River System. Ayeyarwady*

Basin Exploratory Scoping Study (BESS). Final Report; eWater Ltd. (Australian Water Partnership): Canberra, Australia, 2019; Available online: https://www.airbm.org/wp-content/uploads/2019/03/Ayeyarwady-Basin-Exploratory-Scoping-Study-BESS-Final-Report.pdf (accessed on 25 November 2019).

4. Foran, T.; Almeida, A.; Penton, D.; Shrestha, M. Participatory river basin planning for water resource management in Kamala Basin, Nepal. In Proceedings of the International Commission on Irrigation and Drainage (ICID), 8th Asian Regional Conference, Kathmandu, Nepal, 2–4 May 2018; Available online: http://www.icid.org/8arc_postproceedings.pdf (accessed on 24 November 2019).
5. Foran, T.; Manorom, K. Pak Mun Dam: Perpetually Contested. In *Contested Waterscapes in the Mekong Region: Hydropower, Livelihoods and Governance*; Molle, F., Foran, T., Käkönen, M., Eds.; Earthscan: London, UK, 2009; pp. 55–80.
6. Dixit, A.; Gyawali, D. Nepal's constructive dialogue on dams and development. *Water Altern.* **2010**, *3*, 106.
7. Wolsink, M. River basin approach and integrated water management: Governance pitfalls for the Dutch Space-Water-Adjustment Management Principle. *Geoforum* **2006**, *37*, 473–487. [CrossRef]
8. Bergsma, E. Expert-influence in adapting flood governance: An institutional analysis of the spatial turns in the United States and the Netherlands. *J. Inst. Econ.* **2017**, *14*, 449–471. [CrossRef]
9. The Dublin Statement on Water and Sustainable Development. In Proceedings of the International Conference on Water and Environment (ICWE), Dublin, Ireland, 26–31 January 1992.
10. Molle, F. Nirvana concepts, storylines and policy models: Insights from the water sector. *Water Altern.* **2008**, *1*, 131–156.
11. Medema, W.; McIntosh, B.S.; Jeffrey, P.J. From Premise to Practice: A Critical Assessment of Integrated Water Resources Management and Adaptive Management Approaches in the Water Sector. *Ecol. Soc.* **2009**, *13*, 29. [CrossRef]
12. van Koppen, B. Winners and Losers of IWRM in Tanzania. *Water Altern.* **2016**, *9*, 588–607.
13. Clement, F.; Suhardiman, D.; Bharati, L. IWRM discourses, institutional Holy Grail and water justice in Nepal. *Water Altern.* **2017**, *10*, 870–887.
14. Biswas, A.K.; Varis, O.; Tortajada, C. *Integrated Water Resources Management in South and South-East Asia*; Oxford University Press: New Deli, India, 2005.
15. Rautanen, S.-L.; van Koppen, B.; Wagle, N. Community-driven multiple use water services: Lessons learned by the rural Village water resources management project in Nepal. *Water Altern.* **2014**, *7*, 160–177.
16. Lankford, B.A. *From Integrated to Expedient: An Adaptive Framework for River Basin Management in Developing Countries*; International Water Management Institute: Colombo, Sri Lanka, 2007.
17. Union of Myanmar. *Request for Proposal for the Services for C1.17—Development of the Ayeyarwady Decision Support System and Basin Master Plan*; Directorate of Water Resources and Improvement of River Systems, Project Management Unit, Union of Myanmar: Naypyitaw, Myanmar, 2017; (unpublished document).
18. Foran, T. Large hydropower and legitimacy: A policy regime analysis, applied to Myanmar. *Energy Policy* **2017**, *110*, 619–630. [CrossRef]
19. Rasul, G. Beyond hydropower: Towards an integrated solution for water, energy and food security in South Asia. *Int. J. Water Resour. Dev.* **2019**, *62*, 1–25. [CrossRef]
20. Suhardiman, D. The politics of river basin planning and state transformation processes in Nepal. *Geoforum* **2018**, *96*, 70–76. [CrossRef]
21. Molle, F.; Foran, T.; Käkönen, M. *Contested Waterscapes in the Mekong Region: Hydropower, Livelihoods, and Governance*; Earthscan: London, UK, 2009.
22. Conca, K. *Expert Networks: The Elusive Quest for Integrated Water Resources Management, in Governing Water: Contentious Transnational Politics and Global Institution Building*; Conca, K., Ed.; MIT Press: Cambridge, MA, USA, 2006; pp. 123–165.
23. Foran, T. *Rivers of Contention: Pak Mun Dam, Electricity Planning, and State–Society Relations in Thailand, 1932–2004*; University of Sydney: Sydney, Austrilia, 2006.
24. McCaughey, D.; Bruning, N.S. Rationality versus reality: The challenges of evidence-based decision making for health policy makers. *Implement. Sci.* **2010**, *5*, 39–59. [CrossRef]
25. Zey, M. *Criticisms of Rational Choice Models, in Decision Making: Alternatives to Rational Choice Models*; Sage Publications, Inc: Thousand Oaks, CA, USA, 1992; pp. 9–31.
26. Stokey, E.; Zeckhauser, R. *A Primer for Policy Analysis*; W. W. Norton & Company: New York, NY, USA, 1978.

27. Lepenies, R. Discovering the Political Implications of Coproduction in Water Governance. *Water* **2018**, *10*, 1475. [CrossRef]
28. Edelenbos, J.; van Buuren, A.; van Schie, N. Co-producing knowledge: Joint knowledge production between experts, bureaucrats and stakeholders in Dutch water management projects. *Environ. Sci. Policy* **2011**, *14*, 675–684. [CrossRef]
29. Wyborn, C.A. Connecting knowledge with action through coproductive capacities: Adaptive governance and connectivity conservation. *Ecol. Soc.* **2015**, *20*, 11. [CrossRef]
30. Tress, G.; Tress, B.; Fry, G. Clarifying Integrative Research Concepts in Landscape Ecology. *Landsc. Ecol.* **2005**, *20*, 479–493. [CrossRef]
31. IFC. *Strategic Environmental Assessment of the Hydropower Sector in Myanmar. Final Report*; International Finance Corporation (IFC): Washington, DC, USA, 2018.
32. Arcadis. *Integrated Ayeyarwady Delta Strategy. Delta Report. Final version (2.2)*; Arcadis: Yangon, Myanmar, 2018.
33. HIC. *Ayeyarwady State of the Basin Assessment (SOBA) 2017: Synthesis Report, Volume 1, Yangon, December 2017*; Hydro-Informatics Centre: Yangon, Myanmar, 2017.
34. Transparency International Nepal. *A Study of Nepal's Hydro Power Sector*; Transparency International Nepal: Kathmandu, Nepal, 2016.
35. Shrestha, R.S. Hydropower Development: Before and After 1992. *Hydro Nepal J. Water Energy Environ.* **2016**, *18*, 16–21. [CrossRef]
36. WECS. *Section 7. Terms of Reference [Power Sector Reform and Sustainable Hydropower Development Project and Strategic Environmental and Social Assessment]*; Water and Energy Commission Secretariat: Kathmandu, Nepal, 2016.
37. OUP. Oxford English Dictionary. In *OED Online*; Oxford University Press: Oxford, UK, 2000.
38. Fairclough, I.; Fairclough, N. *Political Discourse Analysis: A Method for Advanced Students*; Routledge: Oxon, UK, 2012.
39. Crawford, S.E.S.; Ostrom, E. A Grammar of Institutions. *Am. Political Sci. Rev.* **1995**, *89*, 582–600. [CrossRef]
40. Basurto, X. A Systematic Approach to Institutional Analysis: Applying Crawford and Ostrom's Grammar. *Political Res. Q.* **2010**, *63*, 523–537. [CrossRef]
41. Hurlbert, M.A.; Gupta, J. An institutional analysis method for identifying policy instruments facilitating the adaptive governance of drought. *Environ. Sci. Policy* **2019**, *93*, 221–231. [CrossRef]
42. Neef, A. Transforming Rural Water Governance: Towards Deliberative and Polycentric Models? *Water Altern.* **2009**, *2*, 53–60.
43. Burgess, J. Deliberative mapping: A novel analytic-deliberative methodology to support contested science-policy decisions. *Public Underst. Sci.* **2007**, *16*, 299–322. [CrossRef]
44. Gastil, J.; Levine, P. *The Deliberative Democracy Handbook: Strategies for Effective Civic Engagement in the Twenty-First Century*; Jossey-Bass: San Francisco, CA, USA, 2005.
45. Lane, S.N. Doing flood risk science differently: An experiment in radical scientific method. *Trans. Inst. Br. Geogr.* **2011**, *36*, 15–36. [CrossRef]
46. Carr, G. Stakeholder and public participation in river basin management—An introduction. *Wiley Interdiscip. Rev. Water* **2015**, *2*, 393–405. [CrossRef]
47. Straton, A.T. Exploring and Evaluating Scenarios for a River Catchment in Northern Australia Using Scenario Development, Multi-criteria Analysis and a Deliberative Process as a Tool for Water Planning. *Water Resour. Manag.* **2011**, *25*, 141–164. [CrossRef]
48. Hajkowicz, S.; Collins, K. A Review of Multiple Criteria Analysis for Water Resource Planning and Management. *Water Resour. Manag.* **2007**, *21*, 1553–1566. [CrossRef]
49. Emerson, K.; Nabatchi, T.; Balogh, S. An Integrative Framework for Collaborative Governance. *J. Public Adm. Res. Theory* **2012**, *22*, 1–29. [CrossRef]
50. Alford, J.; Head, B.W. Wicked and less wicked problems: A typology and a contingency framework. *Policy Soc.* **2017**, *36*, 397–413. [CrossRef]
51. Pahl-Wostl, C. *Water Governance in the Face of Global Change: From Understanding to Transformation*; Springer International Publishing: Cham, Switzerland, 2015.

52. Pahl-Wostl, C.; Knieper, C. The capacity of water governance to deal with the climate change adaptation challenge: Using fuzzy set Qualitative Comparative Analysis to distinguish between polycentric, fragmented and centralized regimes. *Glob. Environ. Chang.* **2014**, *29*, 139–154. [CrossRef]
53. AWP. *Activity Plan. Activity 5: Ayeyarwady Exploratory Basin Scoping Exercise*; Australian Water Partnership: Canberra, Australia, 2017; (unpublished document).
54. Costanza, R. *Planning Approaches for Water Resources Development in the Lower Mekong Basin*; Portland State University, Mae Fah Luang University: Portland, OR, USA, 2011.
55. Suhardiman, D.; Clement, F.; Bharati, L. Integrated water resources management in Nepal: Key stakeholders' perceptions and lessons learned. *Int. J. Water Resour. Dev.* **2015**, *31*, 284–300. [CrossRef]
56. Jalsrot Vikas Sanstha and Policy Entrepreneurs Inc. *Formulation of Basin Development Goals and Pathways: For Delivery of Stakeholder Participation and Strategy Development in the Kamala River Basin, Nepal*; Jalsrot Vikas Sanstha, Policy Entrepreneurs Inc.: Kathmandu, Nepal, 2018.
57. JICA. *Master Plan Study on the Kosi River Water Resources Development. Final Report. March 1985*; Japan International Cooperation Agency: Tokyo, Japan, 1985. Available online: http://open_jicareport.jica.go.jp/pdf/10313831_01.pdf (accessed on 24 November 2019).
58. JICA. *Preparatory survey on JICA's Cooperation Program for Agriculture and Rural Development in Nepal—Food Production and Agriculture in Terai—Final Report*; Japan International Cooperation Agency: Tokyo, Japan, 2013. Available online: http://open_jicareport.jica.go.jp/pdf/12127288.pdf (accessed on 24 November 2019).
59. Montibeller, G. Behavioral Challenges in Policy Analysis with Conflicting Objectives. In *Recent Advances in Optimization and Modeling of Contemporary Problems*; Institute for Operations Research and the Management Sciences (INFORMS): Catonsville, MD, USA, 2018; pp. 85–108.
60. CSIRO; WECS. *State of the Water Resources in the Kamala Basin, Nepal*; CSIRO Land and Water: Canberra, Australia. (in press)
61. Carson, L.; Hartz-Karp, J. Adapting and Combining Deliberative Designs. Juries, Polls, and Forums. In *The Deliberative Democracy Handbook: Strategies for Effective Civic Engagement in the Twenty-First, Century*; Gastil, J., Levine, P., Eds.; Jossey-Bass: San Francisco, CA, USA, 2005.
62. Tan, P.-L.; Bowmer, K.H.; Mackenzie, J. Deliberative tools for meeting the challenges of water planning in Australia. *J. Hydrol.* **2012**, *474*, 2–10. [CrossRef]
63. Hughey, K.F.D.; Jacobson, C.L.; Smith, E.F. A framework for comparing collaborative management of Australian and New Zealand water resources. *Ecol. Soc.* **2017**, *22*, 28. [CrossRef]
64. Benson, D.; Jordan, A.; Smith, L. Is Environmental Management Really More Collaborative? A Comparative Analysis of Putative 'Paradigm Shifts' in Europe, Australia, and the United States. *Environ. Plan. A Econ. Space* **2013**, *45*, 1695–1712. [CrossRef]
65. Maclean, K.; Cullen, L. Research methodologies for the co-production of knowledge for environmental management in Australia. *J. R. Soc. NZL* **2009**, *39*, 205–208. [CrossRef]
66. Pahl-Wostl, C. The role of governance modes and meta-governance in the transformation towards sustainable water governance. *Environ. Sci. Policy* **2019**, *91*, 6–16. [CrossRef]
67. Alston, M. Water policy, trust and governance in the Murray-Darling Basin. *Aust. Geogr.* **2016**, *47*, 49–64. [CrossRef]

Article

Participatory Modelling of Surface and Groundwater to Support Strategic Planning in the Ganga Basin in India

Marnix van der Vat [1,*], Pascal Boderie [1], Kees C. A. Bons [1], Mark Hegnauer [1], Gerrit Hendriksen [1], Mijke van Oorschot [1], Bouke Ottow [1], Frans Roelofsen [1], R. N. Sankhua [2], S. K. Sinha [3], Andrew Warren [1] and William Young [4]

1 Deltares, 2600MH Delft, The Netherlands; pascal.boderie@deltares.nl (P.B.); kees.bons@deltares.nl (K.C.A.B.); mark.hegnauer@deltares.nl (M.K.); gerrit.hendriksen@deltares.nl (G.H.); mijke.vanoorschot@deltares.nl (M.O.); bouke.ottow@deltares.nl (B.O.); frans.roelofsen@deltares.nl (F.R.); andrew.warren@deltares.nl (A.W.)
2 Central Water Commission, New Delhi 110066, India; sankhua12@yahoo.com
3 Central Ground Water Board, New Delhi 110011, India; sujitsinha95@gmail.com
4 World Bank, Washington, DC 20433, USA; wyoung@worldbank.org
* Correspondence: marnix.vandervat@deltares.nl

Received: 13 August 2019; Accepted: 18 November 2019; Published: 21 November 2019

Abstract: The Ganga Basin in India experiences problems related to water availability, water quality and ecological degradation because of over-abstraction of surface and groundwater, the presence of various hydraulic infrastructure, discharge of untreated sewage water, and other point and non-point source pollution. The basin is experiencing rapid socio-economic development that will increase both the demand for water and pollution load. Climate change adds to the uncertainty and future variability of water availability. To support strategic planning for the Ganga Basin by the Indian Ministry of Water Resources, River Development and Ganga Rejuvenation and the governments of the concerned Indian states, a river basin model was developed that integrates hydrology, geohydrology, water resources management, water quality and ecology. The model was developed with the involvement of key basin stakeholders across central and state governments. No previous models of the Ganga Basin integrate all these aspects, and this is the first time that a participatory approach was applied for the development of a Ganga Basin model. The model was applied to assess the impact of future socio-economic and climate change scenarios and management strategies. The results suggest that the impact of socio-economic development will far exceed the impacts of climate change. To balance the use of surface and groundwater to support sustained economic growth and an ecologically healthy river, it is necessary to combine investments in wastewater treatment and reservoir capacity with interventions that reduce water demand, especially for irrigation, and that increase dry season river flow. An important option for further investigation is the greater use of alluvial aquifers for temporary water storage.

Keywords: integrated water resources management; river basin planning; Ganga River; India; participatory modelling; conjunctive water use; hydrologic modelling

1. Introduction

The Ganga River Basin (Figure 1) in India stretches over 860,000 km^2 [1] and is home to more than 485 million people (2011 census data [2]). The population is concentrated on the plains that support extensive irrigated agriculture. The plains are of very low slope, falling from 250 m above mean sea level in the west, to approximately 25 m near Farakka at the border with Bangladesh—a distance of over 1500 km. North of the plains the Ganga and its tributaries flow from the Himalaya at elevations over 6000 m. Covered by snow and glaciers, the Himalaya significantly influence the flow regime in the northern tributaries. The mountains and hills to the south are much lower, with an average elevation of around 1000 m. Water availability increases in the plains from west to east. The Himalayan tributaries of the Ganga (the Yamuna, Ghagra, Gandak and Kosi) supply the majority of the water to the plains. Conjunctive irrigation using surface and groundwater in the western part of the plains has led to local decreases in groundwater tables, while in some canal and eastern areas waterlogging is a major problem. In the basin, precipitation increases further to the east, as does mainstem flow as tributaries join. Pre-monsoon water shortage is common in dry years, especially in the western plains.

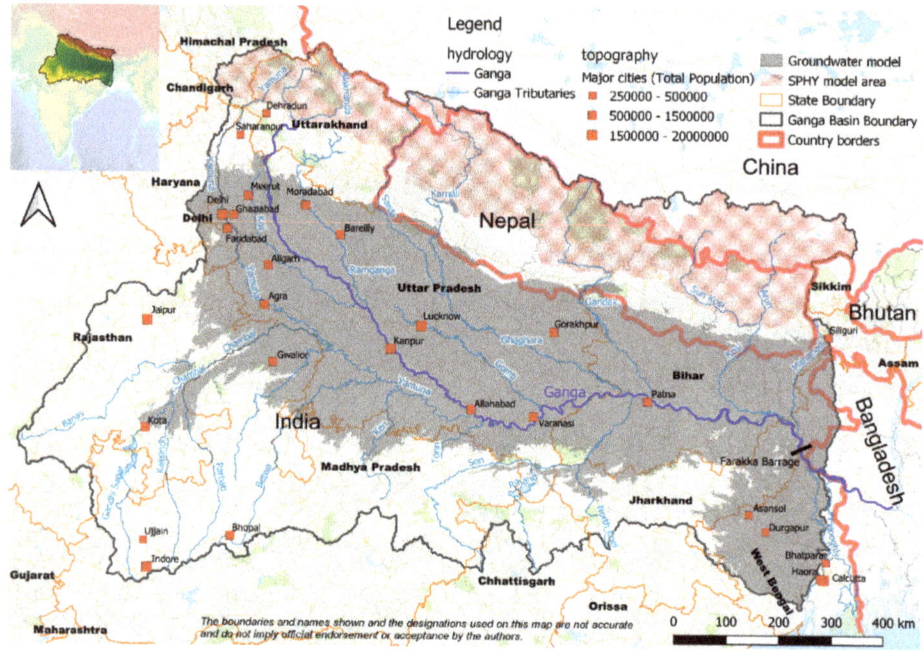

Figure 1. Ganga river basin map. Background based on Wikimedia unlabeled layer.

Water is diverted from rivers through canals and pumped from groundwater. A large fraction of irrigation water is not used for plant transpiration and returns as aquifer recharge or drainage to canals and rivers. There are direct exchanges between the rivers and groundwater. Depending on river and groundwater level, the flow is either from groundwater to river (gaining river) or from rivers/canals to groundwater (losing river). Water quality and riverine ecology depend strongly on the flows resulting from the interaction between geo-hydrology and water resources management.

The ecological health of the Ganga River and some of its tributaries has deteriorated significantly due to high pollution loads from point and non-point sources; river modifications with infrastructure (dams and barrages); flow regime changes caused by high levels of water abstraction, mostly for irrigation, but also for municipal and industrial uses; and hydropower generation [3]. The Government of India has committed to an ambitious goal of rejuvenating the Ganga and has assigned significant funds to address the problem [4]. Since India is a federated country, and responsibility for water resources management is assigned to the states by the Constitution cooperation with and between the national government and those of the 11 Indian states is required for effective basin management.

The Ganga River Basin Model was developed by a collaborative team of national and international scientists with funding from the South Asia Water Initiative (a multi-donor trust fund managed by the World Bank) to support strategic river basin planning. It assesses the impacts of different socio-economic and climate change scenarios combined with different strategies for new infrastructure, management and operation. The objective of applying a participatory approach to model development was to both improve the quality of the model and to increase the commitment and ownership of relevant authorities and agencies. The process of model construction and the assessment of the first scenarios and strategies led by international scientists are intended as the start of a continuous process of model application and improvement led by Indian authorities and agencies. A set of reports provides a description of the set-up and calibration of the Ganga River Basin Model [5], a description of the participatory modelling process [6], and presentation and discussion of the scenario modelling results, environmental flow analysis, and surface-groundwater analysis [7].

2. Materials and Methods

2.1. Participatory Modelling

The technical complexity and scale of the Ganga Basin makes its rejuvenation a problem in which there is a need both for enhanced system understanding and for balancing of a diversity of stakeholder values and perspectives. A participatory modelling approach [8] was applied to facilitate a robust technical analysis to increase existing knowledge on the Ganges River system. Direct interaction with stakeholders facilitated the input of important local knowledge, open discussion of results, interventions, scenarios and strategies. This is the first time a participatory modelling approach has been applied for the Ganga Basin in India.

Participatory modeling refers in this case to the integration of four distinct approaches that can be applied to support strategic basin planning: (i) water resources planning, e.g., the assessment of the impact of different planning alternatives; (ii) the use of scientific knowledge by means of computer-based models to assess impacts; (iii) stakeholder participation in the definition of objectives, indicators, models, interventions, scenarios and strategies; and (iv) collaboration, in the sense of negotiation between stakeholders to reach a decision on the desired plan (Figure 2) [8].

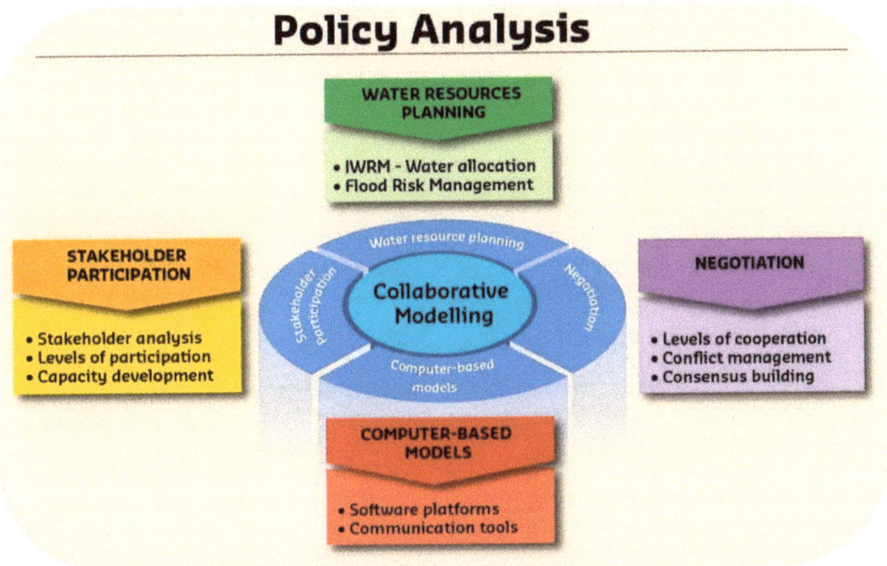

Figure 2. Collaborative modeling for policy analysis.

For system and model definition we adapted and applied a Group Model Building (GMB) approach, a participatory technique to explore constructively and synthesize the multitude of stakeholder perceptions of the interactions in the river system [9,10]. GMB was used particularly for problem and indicator identification and scenario definition, and its results informed the development and application of the computational framework. Further stakeholder engagement was carried out for model validation and strategy development in working groups and workshop settings, where a combination of plenary and focus group discussion techniques were applied.

To effectively involve the range of key technical partners and other stakeholders (both hereafter collectively referred to as 'stakeholders') in the entire Ganga Basin in India, we distinguished different geographical levels (basin and state), and different involvement levels (circles of influence).

2.1.1. Determining Stakeholder Involvement: 'Circles of Influence'

We used a circles of influence approach [11] to structure the participatory planning process for the Strategic Basin Planning in the Ganges River Basin. This approach has been successfully applied in many programs and projects worldwide.

The circles of influence approach engages different stakeholders in various formats and levels of intensity. The generic Circles of Influence framework includes four circles: Circle A—Model Developers, Circle B—Model Users and Validators, Circle C—Interested Parties, and Circle D—Decision Makers (Figure 3). In this project, Circle A stakeholders comprised governmental agencies, such as CWC (Central Water Commission) and CGWB (Central Ground Water Board), and knowledge institutes, such as NIH (National Institute of Hydrology). They were responsible for co-developing the model together with the international technical team. They were trained in hydrologic and river system modelling and were involved in training and capacity building of stakeholders in other circles. Several working groups were organized with Circle B stakeholders—the model validators and users. Circle C stakeholders were consulted at several moments throughout the planning process through multiple consultation meetings conducted in each of the riparian states. Decision makers (Circle D) were also periodically informed and consulted. Stakeholder identification and analysis was conducted during project inception to map stakeholders to their respective circles of influence [11].

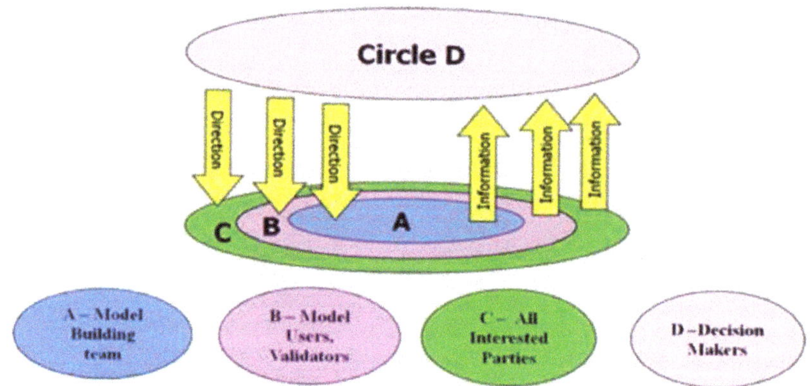

Figure 3. The "circles of influence" process framework [11].

2.1.2. Implementation of the Participation Process

The different stakeholder groups were each engaged in different ways at different stages of the development of the decision-support system (Table 1). The dashboard mentioned in the table is a visualization tool developed to analyze and compare model results (see Section 2.2.6 for details).

Table 1. Summary of steps in stakeholder engagement (A = Model Developers; B = Model Users; C = Interested Parties; D = Decision Makers).

Phase/Step		Stakeholder Groups	Activity
		Conceptualization Phase	
1	Definition of methods, models and integration	A, B	Several meetings at Delhi and Roorkee
		A, B, D	29 January 2016, first basin-wide Workshop
2	Data Collection and Analysis	A, B	Small working group meetings per topic
		Collaborative Modelling Phase	
3	Identification of indicators	D and partly C	February–May 2016, first series of meetings in all eleven states
		B, D	Questionnaires from stakeholder organizations in the eleven states
4	Model set-up (indicators, schematization, assumptions)	A, B, D	July 2016, second basin-level workshop
		A, B, C, D	July–November 2016, second series of workshops in all eleven states
5	Model calibration	A, B	Small working group meetings by topic
6	Draft version dashboard	A, B	
		Scenario Building Phase	
7	Scenario definition	A	Small working group meetings by task
8	Selection of indicators	A	
9	Model application current situation	A	
10	Model application scenarios and strategies	A	
11	Strategy development (packages of measures, dashboard)	A, B, C, D	March 2017, third basin-level workshop
12	Final version dashboard	A, B, C, D	March–June 2017, third series of workshops in all eleven states
		Consolidation Phase	
13	Training and dissemination	A, B	Small working group meetings by task
14	Presentation of realistic scenarios and strategies		January 2018, fourth basin-level workshop

In the basin-wide workshops, representatives from national organizations as well as from the eleven States participated. In the different state workshops, participation was limited mostly to representatives from organizations from the hosting state. The first basin-wide workshop (January 2016) and the first series of state meetings in all eleven Ganga Basin States (February–May 2016) were used to introduce project assignments to state-level stakeholders and seek stakeholder responses on the project assignment and circulate stakeholder questionnaires to assess concerns and ideas on water management in the Ganga Basin. The second basin-wide workshop (July 2016) and the second series of state meetings (July–November 2016) were used to validate and further elaborate the findings from the questionnaires for input into the technical modeling process. On average, 25 participants from 10–15 organizations participated in each of the state workshops.

The results of the second series of consultations was used to improve the Ganga River Basin Model and the dashboard developed to present its results. The third basin-wide workshop (March

2017) and the third series of state meetings (March–June 2017) focused on the validation of the model results for the present situation and for the development of scenarios and strategies.

2.2. Computational Framework

A model to support strategic planning should include all essential components of the system and their interactions in order to be able to assess the impact of scenarios and strategies. However, the amount of detail that can be included in the model is limited. The value of the model is in its schematic representation of reality. Previous modelling exercises for the Ganga Basin include:

1. Water systems modelling for Ganga Basin by INRM Consultants Pvt. Ltd. [12], which applied the SWAT model;
2. Ganges river basin modelling by Institute of Water Modelling [13], which applied the MIKE BASIN model; and
3. Surface and groundwater modelling of the Ganga River Basin by IIT (Indian Institutes of Technology) [14], which applied SWAT and MODFLOW.

All three reports mention issues regarding lack of data for model input as well as calibration, and all three show calibration results with varying degrees of acceptability, as well as results from limited scenario analysis. Ref. [12,13] are limited to surface water hydrology. Ref. [14] combines modeling of surface and groundwater, but the interaction between the two systems is not modelled dynamically. Only [12] includes water quality modeling, but no calibration is included. None of these modelling exercises includes impacts on ecology and the results. The recommendations of these studies have had little impact on management and planning for the Ganga Basin [15]. Model input and output generated as part of these three studies have not been made available publicly and can therefore not be used as a starting point for new modeling.

The Ganga River Basin Model presented in this paper has a very wide scope allowing for an integrated assessment of impacts related to hydrology, geohydrology, water resources management, water quality and ecology. This is the first time that all these aspects have been integrated into one modelling approach. The level of detail included is limited to keep the model manageable and the complexity understandable.

The model components and their interactions (Figure 4) are described in the remainder of this Section, focusing on the input data and the interaction between the model components with references provided for detailed descriptions of the individual components.

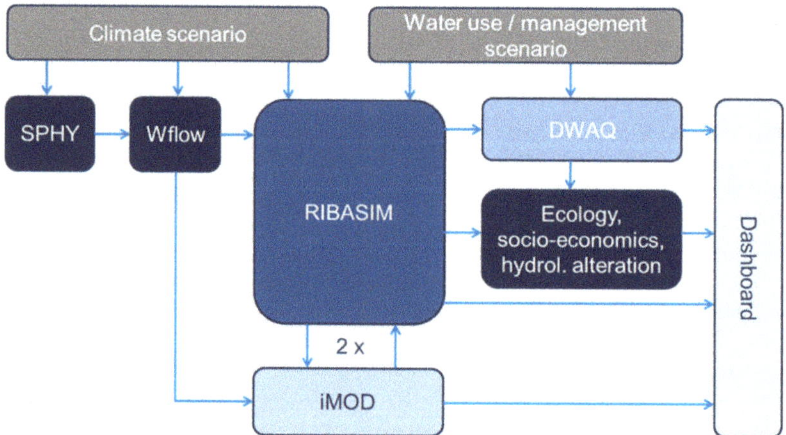

Figure 4. Schematic representation of the workflow of the different model components of the Ganga River Basin Model.

The hydrological models cover the entire Ganga Basin upstream of Farakka Barrage including those parts of the upstream basin located in Nepal and China. This permits robust assessment of the upstream flows. On request of the state of West Bengal, the part of the catchment west of the Hooghly branch below Farakka was also been included in the model area. The remainder of the analysis focuses on the Indian part of the basin upstream of Farakka Barrage.

2.2.1. Hydrological Models

The description of basin hydrology uses the SPHY [16,17] and Wflow [18] models. These are fully distributed models implemented on a 1 km by 1 km grid. SPHY describes the hydrological process in the mountainous areas of the Himalaya and was selected as it is specifically designed for glacier and snow hydrology and it has previously been successfully applied to the Himalaya [19]. Rainfall-runoff for the non-mountainous part of the Ganga Basin were simulated with Wflow—a general-purpose hydrological model. River discharges from SPHY provide inputs to Wflow. Both models use the following static input data:

- A digital elevation model (DEM) derived from the HydroSheds SRTM DEM [20]
- A shapefile of the main rivers derived from Open Street Map [21]
- Land use/land cover map for the Indian part of the model area from the Indian Institutes of Technology (IIT), based on data from the National Remote Sensing Centre (NRSC) [22]) and the GlobCover map [23] for the parts in Nepal and China
- A soil map based on FAO's Soil Map of the World [24] and a soil map with quantitative soil properties for the topsoil and subsoil [25]

SPHY also uses a map of glacier outlines and distinction between debris-covered and debris-free glacier surfaces from [26].

Both models use the following distributed meteorological data:

- Precipitation inside India for the period 1959–2012 from the Indian Meteorological Department (IMD) [27] and for the years 2013 and 2014 data from the WFDEI data set [28]. Outside India the EUWATCH dataset [29] were used for the period 1959–1978 and WFDEI for 1979–2014.
- For temperature and potential evapotranspiration, the IMD data could not be used due to an issue with the interpolation in the Himalayas [30]. Therefore, the global data sets from EUWATCH, for 1959–1978, and WFDEI, for 1979–2014, were used for the entire model domain.

The concepts used by SPHY and Wflow to simulate river flow are described in [16,17] and [18] respectively. They produce gridded, daily flows across the entire model domain, which are input to the water resources model (see Section 2.2.2), and gridded, daily infiltration rates, which are input to the groundwater model (see Section 2.2.4).

2.2.2. Water Resources Model

RIBASIM [31,32] simulates the use and distribution of water using river discharges from Wflow as input. It uses a schematization of links and nodes (Figure 5) to describe the flow of water in the rivers, storage in reservoirs, diversions into canals, and consumptive use and return flows across the basin. Water can be used from precipitation, rivers, canals, or from groundwater. Conjunctive use of surface and groundwater is included. Return flows can be divided between rivers, canals and groundwater. This is important for the Ganga plains, where extensive leakage from irrigation canals recharges groundwater aquifers. RIBASIM was linked to the groundwater model by simulation of extraction and infiltration rates and by the flux between the river and the groundwater, as simulated by the groundwater model.

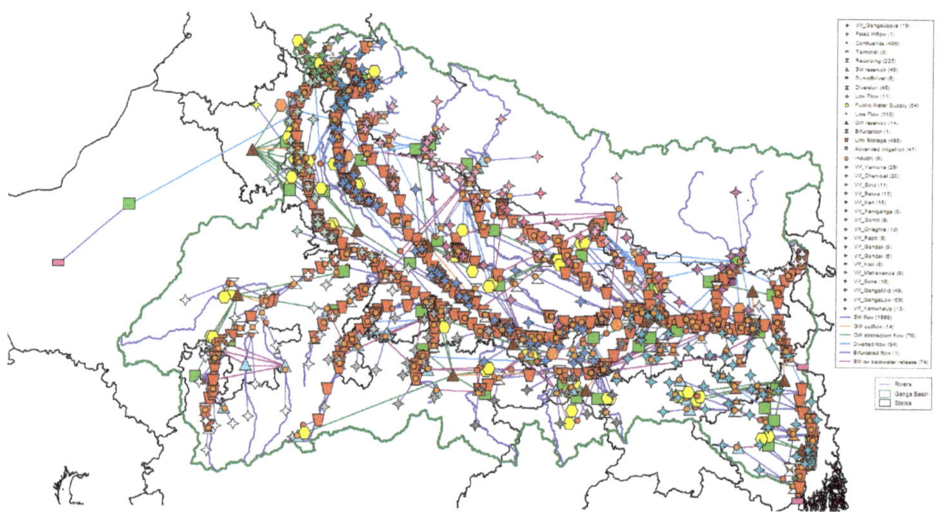

Figure 5. Schematization of the Ganga River Basin in RIBASIM.

Data for water infrastructure such as barrages, dams and canals were mainly been derived from the Ganga Basin Report [33] and the India-WRIS (Water Resources Information System [34]). The schematization was adapted with input from the first round of state and basin-wide workshops, including on the location of existing and planned reservoirs and canals, the compartmentalization of irrigated areas in command areas, and the main abstractions of surface and groundwater.

Information on irrigated crop areas was derived from the Land Use Statistics Information System of the Ministry of Agriculture and Farmers Welfare [35]. Data for 246 districts was aggregated to data for 41 irrigation nodes. The cropping calendar, describing when which crop is planted, was derived from information provided by the Crop Science Division of the Indian Council of Agricultural Research [36]. Estimates for irrigation efficiencies and return flow fractions from [37] and monthly average reference evapotranspiration data per state from [38] were used. Monthly crop transpiration coefficients for most crops are India specific values from [39], but for maize and rapeseed coefficients from [40] were used. For sugarcane, tobacco and fodder crops no information specific for India could be found, and values from FAO [41] were used.

District-level population data from the 2011 census data [2] were used to assess domestic water demands and extrapolated to 2015 based on projections for 2001 to 2026 [42] and on urbanization rates for the period 2001–2011. District data were aggregated to correspond with the 55 public water supply nodes for domestic demand. Data on water sources, leakage and return flow for major cities from [43] and data on industrial water demand from [44] were also used.

The water demand of public water supply and irrigation nodes can be fulfilled by water from surface and groundwater resources to simulate conjunctive use. The capacity of the surface water supply for irrigation is determined by the canal capacity, mostly obtained from India-WRIS, and for public water supply from [43]. The capacity of groundwater supply for both has been tuned to yield results that are comparable to the estimates presented in [45].

During periods of water shortage, RIBASIM allocates water based on priorities. The following ranking of priorities was used in the Ganga Basin Model:

1. Drinking water supply;
2. Industrial water supply;
3. Irrigation water supply;
4. Low flow requirements for spiritual use, bathing and environmental flows.

The concepts used by RIBASIM to simulate water demand and allocation, and the operation of infrastructure, are described in [31,32]. RIBASIM results include monthly flows in rivers and canals, groundwater abstraction rates, and the water supplied to fulfil each water demand. The simulated groundwater abstraction rates are used as input to groundwater model. Simulated river flows are used in the groundwater model to assess the water level that determines the exchange between the river and the groundwater. In the workflow (Figure 4) the water resources model is run twice: once before the groundwater simulation with zero exchange between rivers and groundwater, and once after the completion of the groundwater simulation with the exchanges as calculated by the groundwater model. The simulated flows in the rivers are used as input to the water-quality model, the ecological assessment module, and the dashboard.

2.2.3. Water-Quality Model

Water quality is assessed using DWAQ [46,47] by combining RIBASIM discharges with pollutant load estimates. DWAQ applies the advection-diffusion equation using a numerical solution based on finite volumes derived from the RIBASIM calculation grid to obtain pollutant concentrations of, among others, BOD5 (Biological Oxygen Demand in 5 days) and coliform bacteria. For RIBASIM, links representing schematized canals without flow volumes are estimated based on length of the link and estimated maximum flow velocity in the link. Decay of BOD5 and coliforms in the Ganges and Yamuna depends on simulated residence time and kinetic rate constants adapted from [48] and adjusted by calibration against surface water quality measurements obtained by CWC and CPCB (Central Pollution Control Board) for the period 1999–2014. In DWAQ, pollutants enter surface water as net emissions representing the non-treated fraction of the total waste load generated at RIBASIM nodes. Gross emissions are the product of emission variables and specific emission factors as follows:

- number of rural and urban population multiplied by waste production per capita [44,49,50];
- industrial effluent [44] multiplied by the typical effluent concentrations for chemicals [49,51], distilleries [52], dying textile and bleaching [49,51], food, dairy and beverages [53,54], pulp and paper [55,56], sugar [57] and for tanneries and others [51,58,59]; and
- cropping pattern area (Section 2.2.2) multiplied by specific emission factors to represent irrigation losses by leaching from soil [59].

Sewage effluent and treatment is modelled separately, considering volumetric treatment capacity based predominantly on [40] and removal efficiency by contaminant from [60,61].

2.2.4. Groundwater Model

Groundwater movement is simulated by iMOD [62], the Deltares extension of the well-known MODFLOW code [63] for solving the groundwater flow equation. iMOD uses the same calculation grid as Wflow, but is applied only to the alluvial area of the basin. It was not possible to model groundwater in the hard-rock areas because of a lack of data on surface-groundwater connectivity. iMOD simulations are transient, while recharge, abstraction and surface water level data inputs are time-dependent.

iMOD describes the alluvial aquifers using geological information. Fence diagrams were available from CGWB as well as MAP files describing the thickness of geologic layers. The result is a three-layer aquifer conceptualization of variable thickness. In the mountainous areas to the south, the shallow aquifer thins; it is thickest in the central basin with a maximum depth of approximately 400 m below sea level. Aquifer parameters (permeability, storage coefficient) were provided by CGWB based on modelling studies by Indian Institutes of Technology [14]. Groundwater recharge was obtained from Wflow (grid-based) for non-irrigated areas and from RIBASIM (lumped) for irrigated areas.

Based on RIBASIM river discharge, river water levels are derived on a 1 km scale and used to calculate fluxes between the river and groundwater. This approach was applied to the Ganga and its main tributaries. For the intermediary areas the surface water system, represented by minor streams and local drainage, is modelled using the MODFLOW Drain Package [63]. This simulates head-dependent flux boundaries, such as the exchange between the groundwater and local surface water.

For each RIBASIM node, groundwater demand for irrigation, industry and public water supply is estimated. For iMOD, all demands besides irrigation are equally distributed as abstraction wells on a 1 km scale over each node area. Irrigation abstraction is spatially distributed using additional information from the irrigation map developed by the International Water Management Institute [64] that indicates irrigation areas and irrigation source. Abstraction wells are only located in cells indicated with irrigation from groundwater.

The CGWB manages a widely-distributed network of nearly 9000 groundwater monitoring locations. Data from this network was made available for calibration. A selection of 1800 locations was used to adjust the model parameters.

2.2.5. Ecological Assessment Module

The ecological assessment module translates the simulated impact of scenarios and interventions on hydrology and water quality into impacts on ecology and ecosystem services. To calculate the overall hydrological indicator, ten ecologically-relevant hydrological sub-indicators were identified to give an indication of changes in magnitude, duration, timing and frequency of both low and high discharge events compared to the pristine condition.

Ecological sub-indicators are expressed as changes in habitat suitability compared to the pristine situation for several fish species, the Ganga river dolphin, the Gharial and the Indian Flapshell turtle. Habitat suitability was calculated with response curves containing environmental thresholds for water quality and water depth.

For socio-economics, the sub-indicators are fisheries, ritual bathing and floodplain agriculture. The fisheries score depends on habitat suitability for the commercially valuable fish species; the religious bathing scope depends on water depth and BOD, coliform bacteria are less important due to the lower risk of contamination during religious bathing when compared to swimming; and the floodplain agriculture score depends on previously flooded bare areas that becomes available during the dry season.

Since the Ganga River and its tributaries differ in geomorphology, discharge and anthropogenic pressures, the system is subdivided into relatively homogeneous eco-zones. Habitat suitability of species is calculated by eco-zone with dose–effect relations for water depth, dissolved oxygen and temperature, which are extracted from the results of the other parts of the Ganga River Basin Model. Ecological scores are calculated as a percentage of agreement with the reference situation, which is the simulated pristine situation without anthropogenic pressures and with historical land use.

The ecological assessment module and its application to the analysis of different flow regimes is described in detail in [7].

2.2.6. Water Information System and Dashboard

All model inputs and all relevant outputs are stored in the database of the water information system GangaWIS (Figure 6). Delft-FEWS [65] is used to run the different model components and Delft-FEWS model connectors are used to export input data from the database to different model components and to import simulation results from the model components into the database. The database consists of a PostgreSQL/PostGIS [66] geodatabase for time series and vector data and a THREDDS server [67] to store and retrieve gridded data in NetCDF format [68].

The dissemination layer has three components. A website [69] presents static information, such as the project description and reports. The Delta Data Viewer [70] is used to present data from the database on a webpage. And the dashboard presents simulation results aggregated into eleven indicators (Table 2) supported by maps and a Ganga River long-section (profile). All indicator values are calculated based on the simulation results of the meteorological time period 1985–2014.

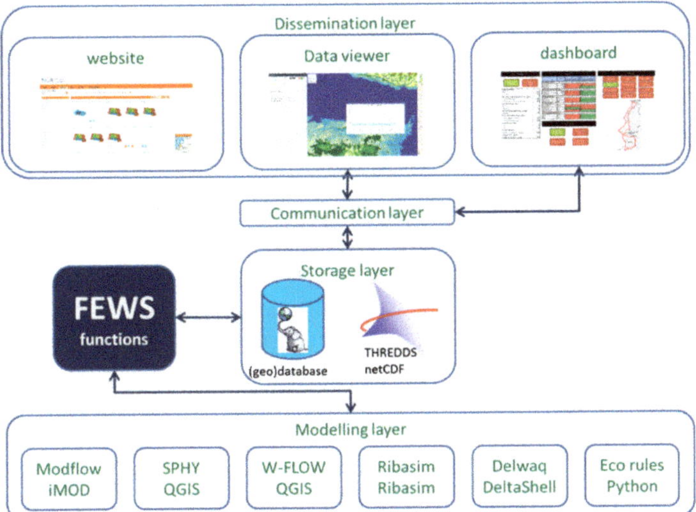

Figure 6. GangaWIS (water information system) structure.

Table 2. Description of the eleven indicators presented on the dashboard of the GangaWIS (see [5] and [7] for detailed descriptions).

Indicator	Description
State of groundwater development (% critical areas)	The percentage of the area where the simulated groundwater abstraction amounts to 90% or more of the simulated recharge. The basis for this information is obtained from the model irrigation nodes.
Lowest discharge at Farakka (m^3/s)	The lowest monthly simulated discharge of the Ganga River above the Farakka Barrier for the one-in-ten dry year.
Volume in reservoirs (billion m^3)	The total sum of simulated water stored in the main basin reservoirs at the end of the monsoon period, October, for the one-in-ten dry year.
Agricultural crop production (% of area harvested)	Ratio between the actual and potential harvested area at basin level for the one-in-ten lowest production year.
Deficit irrigation water (%)	The difference between simulated irrigation water supply and simulated demand as a percentage of the simulated demand for a one-in-ten dry year.
Deficit drinking water (%)	The difference between simulated drinking water supply and simulated demand as a percentage of the simulated demand for a one-in-ten dry year.
Surface water quality index	Dimensionless index based on classification as presented in [44] for the parameters total coliforms, BOD5 and dissolved oxygen.
Volume of GW (groundwater) extracted (billion m^3)	The total simulated volume of groundwater abstracted for public water supply and irrigation during the year with the one-in-ten highest abstraction.
E-flow: Ecological status (%)	Average percentage of agreement with the simulated pristine status for the habitat suitability of nine key species
E-flow: Hydrological status (%)	Average percentage of agreement with the simulated pristine status for ten hydrological discharge indicators representing magnitude, timing, duration and frequency low, average and high flows based on the Indicators of Hydrological Alteration method [71].
E-flow: Socio-Economic status (%)	Average percentage of agreement with the simulated pristine status of the indicators for the three ecosystem services religious bathing, fisheries and floodplain agriculture

Separate pages zoom to state-level and add state-specific indicators, maps and graphs. Indicators of interest were determined through the participatory modelling process. The dashboard was designed for end-users to assess the impact of scenarios and strategies by comparing results between two model runs with different inputs.

3. Results

3.1. Results of Participatory Modelling Process

3.1.1. Broad Participation

One of the results of the adopted approach was broad participation from different national-level government departments/agencies and those from the eleven Ganga states, both in the series of basin-wide workshops as well as in the different state-level workshops (Tables 3–5). Participants were particularly positive about the opportunities the approach offered counterparts in different government agencies to collaborate and share cross-sectoral information relevant to Ganga basin planning. Many reported gaining new insights into the river basin, users' needs and interests and the role modeling can play in the planning process.

Table 3. Participation of national-level and state-level organizations in basin-wide workshops.

	1st Workshop New Delhi 29 January 2016	2nd Workshop Lucknow 18 July 2016	3rd Workshop Kolkata 2 March 2017	4th Workshop New Delhi 20 February 2017	Total	Average
# participants from central agencies	51	21	24	38	134	34
# participants from state government	40	33	21	12	106	27
# of participants from other organizations	12	11	1	16	40	10

Table 4. Participation of stakeholder organizations in state-level participatory modelling workshops, July–October 2016.

State [1]	HP	UK	Har	Del	Raj	MP	UP	Jhar	Chh	Bih	WB	Total	Average
# participants per workshop/state	34	26	27	23	25	27	27	32	19	22	22	284	26
# of different departments/organizations present	11	12	8	7	8	10	12	10	8	10	9	105	10

[1] HP = Himachal Pradesh, UK = Uttarakhand, Har = Haryana, Del = Delhi, Raj = Rajasthan, MP = Madya Pradesh, UP = Uttar Pradesh, Jhar = Jharkhand, Chh = Chhattisgarh, Bih = Bihar, WB = West Bengal.

Table 5. Participation of stakeholder organizations in state-level model validation workshops, April–June 2017.

State	HP	UK	Har	Del	Raj	MP	UP	Jhar	Chh	Bih	WB	Total	Average
# participants per workshop/state	28	30	17	29	28	20	25	38	18	36	26	295	27
# of different departments/organizations present	14	17	10	10	14	10	15	8	12	11	12	133	12

3.1.2. Input to the Dashboard

The interactive workshops in 2016 provided input on the most important problems in the basin and provided data for model development. The workshops paid special attention to identifying indicators relevant for stakeholders. The dashboard is based on the indicators identified through this process. Where indicators were proposed that could not be evaluated using the modeling framework, the participatory process helped to manage expectations.

3.1.3. Use of the Ganga River Basin Model in Workshops

In the third round of workshops in April 2017 participants articulated priority interventions. Group discussions confirmed four potentially effective strategies:

- increasing irrigation efficiency
- limiting groundwater abstraction
- increasing waste water treatment, and
- increasing reservoir volume

Demonstration model runs were carried out for the strategies with updated inputs representing the different strategies.

3.1.4. Opportunities for Follow-Up

Following project conclusion, the dashboard and underlying models provide a foundation for coordinated strategic planning in the Ganga Basin. Key to success will be continued stakeholder engagement. In the future, stakeholder engagement could be expanded to include representatives from a wider range stakeholder organizations and community bodies.

3.2. Calibration and Validation of the Ganga River Basin Model

The model components for hydrology, geohydrology and water resources management have been jointly calibrated and validated, as river flows are influenced by water use and water infrastructure operations. Flows were calibrated using 1995–2009 data and validated against 1985–1994 data. Flow calibration and validation focused on the Ganga mainstream and its main tributaries. Calibration and validation data for iMOD were specified by location not time period. The entire calibration process had six steps:

1. Calibration of SPHY flows at locations in the Himalayan catchments upstream of any significant water demand or water infrastructure;
2. Calibration of Wflow flows at locations in the catchments outside the Himalayas upstream of any significant water demand or water infrastructure;
3. Calibration of iMOD groundwater levels with a fixed river water level;
4. Calibration of pumping capacities for irrigation and public water supply in RIBASIM, using estimates of 2011 annual pumping from CGWB (2014), and where data unavailable using canal capacities, assuming no supply shortages in wetter than average years;
5. Combined calibration of SPHY, Wflow and RIBASIM using measured river discharges assuming no supply shortages in wetter than average years and zero flux between rivers and the groundwater; and
6. Combined calibration of SPHY, Wflow, RIBASIM and iMOD using measured river discharges after incorporation of river-groundwater exchanges from step simulation of iMOD.

A complete description of calibration and validation results as well as sensitivity analysis results are in [5]. Figures 7 and 8 show calibration and validation for monthly flows at two locations on the Ganga River. Observed flow data are from CWC. Flow values are omitted in compliance with the Government of India Water Data Policy for classified data. Figure 7 shows results for Rishikesh, where the Ganga descends from the Himalayas onto the plains. Simulations generally agree well with the measurements but underestimate peak monsoon flows. These peak flows are less important from a water supply perspective, as during the monsoon demands (including to fill storage) are far lower than supply. Figure 8 shows results for Varanasi, the most downstream location on the Ganga for which data were available. Again, simulations match measurements well, but with an overestimation of dry season flows.

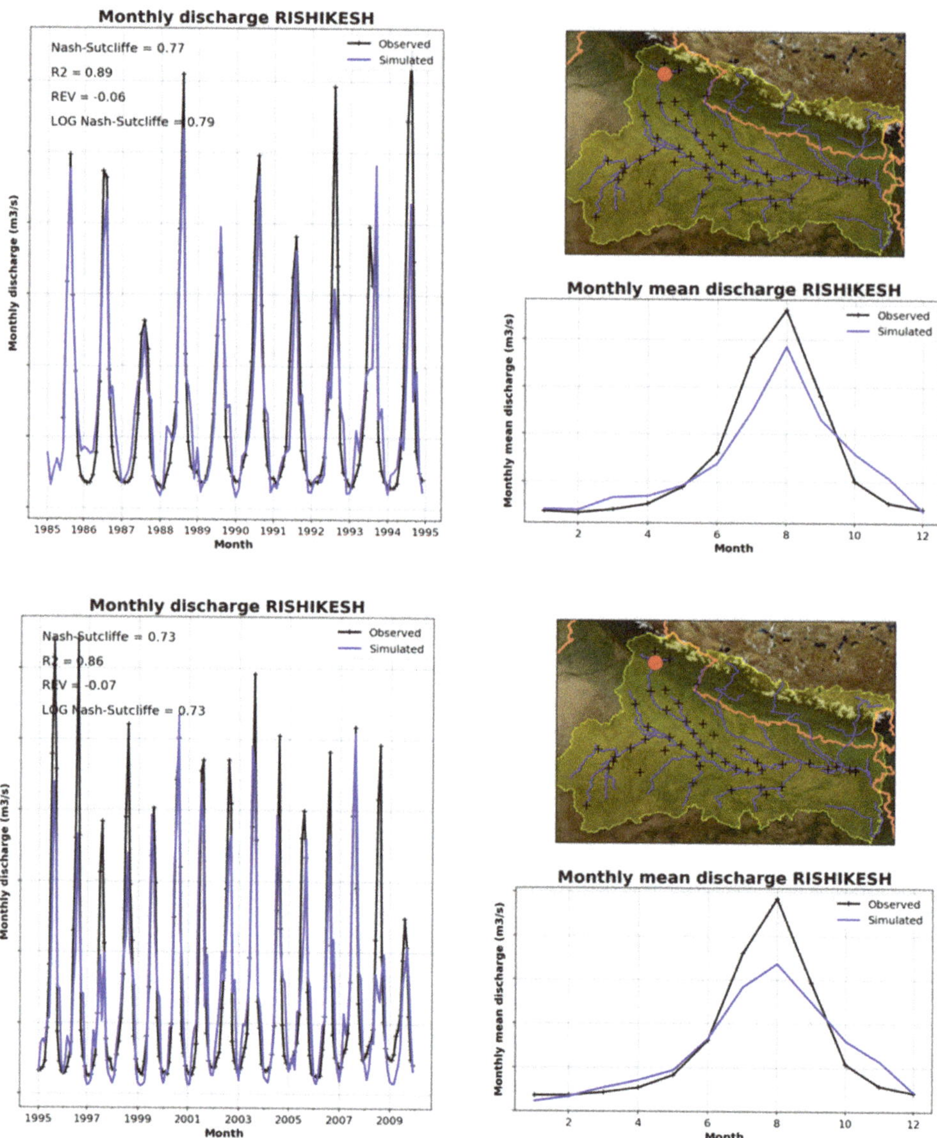

Figure 7. Validation (1985–1994, **top**) and calibration (1995–2009, **bottom**) results for the Ganga River at Rishikesh; monthly discharges (**left**), mean monthly discharges (**right bottom**) and location of the station (red dot on map **right top**).

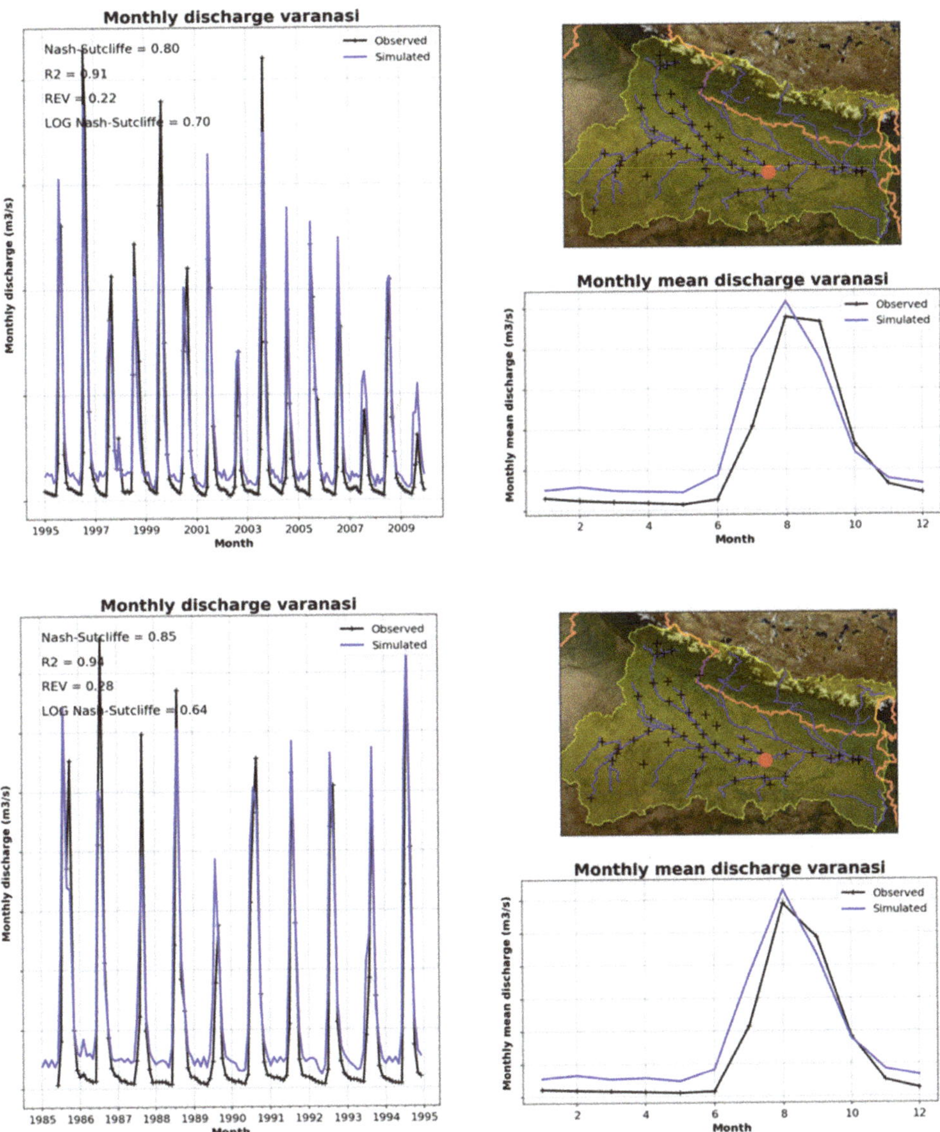

Figure 8. Validation (1985–1994, **top**) and calibration (1995–2009, **bottom**) results for the Ganga River at Varanasi; monthly discharges (**left**), mean monthly discharges (**right bottom**) and location of the station (red dot on map **right top**).

It is difficult to compare the results of model calibration and validation with those of previous studies, since model results of previous studies are only available in the form of reports and since different data were made available to prior studies. Limited availability of measured discharge data within India for the studies reported in [12,13] made these studies focus on stations in Nepal. Ref. [13] reports results for the station Hardinge Bridge in Bangladesh for the period 1998 to 2006. These results show Nash–Sutcliffe efficiency (NSE) coefficients of 0.85 to 0.89, which is comparable to the values presented here for Varanasi (Figure 8), the most downstream station for which data were available in this study. Ref. [14] presents NSE and volume bias for simulation results for 1990 to 2004 for a number of stations within India. For Rishikesh an NSE of 0.60 and a volume bias of +30% is reported, which compares unfavorably with the results presented here where NSE of 0.73 to 0.77 and a bias of −6% to −7% were achieved (Figure 7). There are two more stations both on the Ganga River for which both this study and [14] reports results: Ankinghat and Kanpur. Both studies show comparable values for the NSE, but the volume bias reported in [14] is +30%, while our results vary between −18% to −32%.

Overall, the calibration and validation of this study benefited from better data availability than previous studies. As far as results can be compared, the hydrological results of the Ganga River Basin Model appear to be comparable to the results of previous studies and sometimes represent a slight improvement.

3.3. *Assessment of the Impact of Scenarios and Strategies*

Herein, the term scenario describes developments that impact water resources, but that are outside the direct influence of water managers (e.g., population growth or climate change); and the term strategy describes a combination of interventions designed to address current or future management issues. The effectiveness of strategies can be assessed for different scenarios.

Except for the present scenario, all scenarios are based on assumptions or projections and are, therefore, uncertain. The 'pristine' scenario describes the basin without water resource development. Other scenarios describe possible futures for around the year 2040. All include increases in domestic, industrial, and agricultural water demand. Three climate change futures are considered: no climate change, climate described by the Intergovernmental Panel on Climate Change (IPCC) Representative Concentration Pathway (RCP) 4.5 scenario and climate described by the IPCC RCP8.5 scenario.

Based on stakeholder inputs, strategies were developed that could be implemented in combination or separately:

- **Business as Usual (BAU):** No changes in water resources management.
- **Approved Infrastructure (Appr.Inf)**: Implementation of infrastructure projects approved as of early 2018.
- **Inter-Basin Transfer Links (IBTL+)**: Implementation of the main proposed inter-basin transfers relating to the Ganga Basin.
- **NMCG Planned Treatment (Pl.tr)**: Implementation of the additional treatment plants planned by National Mission for Clean Ganga (NMCG).
- **Improved Treatment (Imp.tr)**: All planned wastewater treatment plants implemented and fully operational, and rural wastewater impact reduced by additional treatment.
- **Increased Irrigation Efficiency (Eff):** Surface water irrigation efficiency increased from 40% to 48% and groundwater irrigation efficiency increased from 70% to 74%.
- **Conjunctive Use (Conj.use)**: Groundwater abstraction reduced by 50% at over-extracted nodes. (Six nodes are over-abstracted in the present scenario and 12 nodes in the 2040 scenarios).
- **E-Flow (e-flow)**: Minimum flow forced to 40% of pristine flow for each month, whenever possible.

Most strategies can be scaled to increase their impact. Figure 9 shows basin-wide indicator values for the modelled scenarios. Impacts are most visible in the hydrological indicators: areas with critical groundwater use increase significantly, and the lowest dry-year river discharge diminishes significantly. The e-flow indicators differ significantly from the pristine condition, however, there are only small differences in e-flow indicators between future scenarios. Scenario assessments (without management interventions) indicate a significant decrease in future water availability, water quality and ecological status. Changes are mainly caused by socio-economic factors, not climate change.

Table 6. Basin wide indicator scores for eight strategies under the IPCC 2040_RCP4.5 scenario.

Indicator	Code	BAU	Appr. Inf	Conj.use	IBTL+	Eff	Pl.tr	Imp.tr	e-flow
State of Groundwater development (% critical areas)	GW over-abstraction	88	88	79	83	88	88	88	95
Lowest discharge at Farakka (m^3/s)	Low Q	1502	1458	1622	1258	1483	1502	1502	1528
Volume in reservoirs (billion m^3)	Res. Store	52	55	53	41	53	52	52	20
Agricultural crop production (% of area harvested)	Agr. Harv.	87	89	74	92	89	87	87	84
Deficit irrigation water (%)	IRR deficit	31	31	47	30	29	31	31	39
Deficit drinking water (%)	DR deficit	34	34	35	35	34	34	34	39
Surface water quality index (-)	WQ index	4	4	4	4	5	4	4	5
Volume of GW extracted (Billion m^3)	GW used	217	215	176	206	207	217	217	235
E-flow: Ecological status (%)	E-ecol	65	65	66	63	66	65	66	73
E-flow: Hydrological status (%)	E-hydr	47	46	49	44	47	47	47	56
E-flow: Socio-Economic status (%)	E-socio	66	66	67	68	66	67	69	75

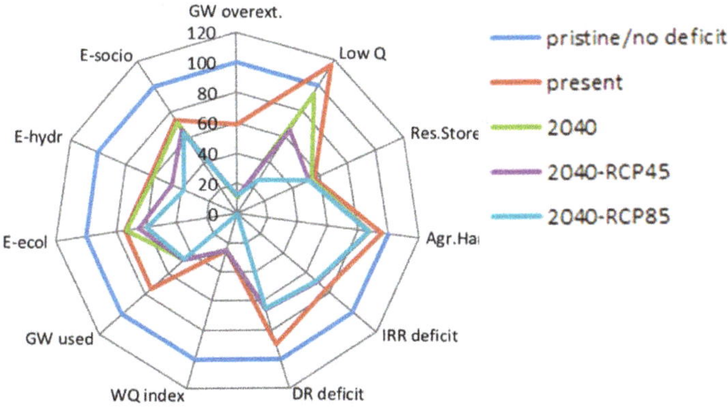

Figure 9. Basin-wide indicators for five scenarios. Values are scaled from 0 to 100, with the pristine scenario as 100). Indicator codes are explained in Table 6 and described in Table 2.

Given the significant potential degradation by 2040, it is informative to evaluate the effectiveness of strategies proposed by stakeholders. Table 6 and Figure 10 show indicator scores for individual strategies under the IPCC 2040_RCP4.5 scenario. If multiple strategies were combined, a greater response would be expected. Details of the assessment are available in [7].

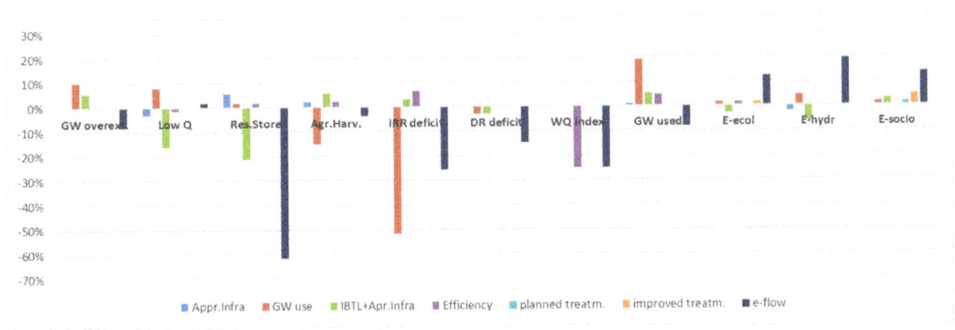

Figure 10. Percent change in basin-wide indicator values for intervention strategies. Values relative to the business as usual scenario. Increases indicate improvement and reduction indicate deterioration.

4. Discussion

The scenario assessments indicate significant degradation from the pristine condition. During part of the year a large fraction of the flow is diverted to canals, sometimes reducing flow in the river to almost zero. Groundwater levels have changed significantly. Flows in the shallow aquifers in the Ganga plains now entirely reflect anthropogenic influences. Water quality has been severely degraded by liquid and solid waste discharges into the river and its tributaries. It is the first time that these findings can be based on comparison of model results for the actual situation with a pristine scenario.

Model results suggest significant additional degradation in water availability, water quality and ecological status will occur in coming decades in the absence of strong management intervention. A new finding from the scenario analysis is that degradation will be mainly caused by socio-economic factors, not climate change. The projected significant increases in water demand by 2040 will mainly affect groundwater, because most available surface river water is already used. Scenario results show that drinking water and irrigation water deficits will increase, and water quality will further deteriorate. Despite considerable data uncertainty, climate change is expected to affect water demand more than water availability.

Results of the analysis show that there is no single simple intervention to address the multiple pressures on the Ganga. A combination of interventions is required. However, the suite of currently considered interventions, which would require huge investment and face significant technical challenges and stakeholder opposition, will not adequately address the future challenges of water availability, water quality and ecology. Indeed, they will not even address the current severe pressures on the river system. This is an important new finding of this study.

The intervention with the greatest potential benefits is further improvement of municipal wastewater treatment. Whether centralized or decentralized, high- or low-technology, greater reduction in pollution improves downstream water quality, improves ecosystem services, and reduces water-related illnesses and deaths. The next most important intervention is an increase in water-use efficiency, especially in irrigated agriculture. Increased efficiency will not immediately increase water availability; however, irrigation deficits may be reduced. This means greater agricultural production for the same level of irrigation withdrawals.

Model results show that water availability in the basin will be insufficient to meet projected future demands and that there are no easy technical solutions. Many interventions that are beneficial for one sector or outcome show negative effects for others.

The results of the scenario analysis show that ambitious strategies are needed to reduce demands across all sectors and that trade-offs need to be made between sectors. The agricultural sector will need to adapt to lower water availability in terms of crop choices, planting seasons and irrigation efficiency. Farmers will need to develop flexible approaches, choosing irrigated or non-irrigated crops depending on monsoon rainfall. This will affect agricultural production and sector employment.

Without coordination and careful balancing of interests, expensive interventions may fail, wasting scarce financial resources. The absence of a functioning water-resources management governance structure in the basin aggravates the challenges the basin is facing. Although not as a result of the presented study, the authors recommend from global experience that a basin management organization with a legal mandate to work across state boundaries is needed to plan the strategy and implement it.

The consequences of the conclusions presented above are far-reaching and will involve many departments and ministries beyond just water resources. Non-technical interventions, including incentives to change cropping patterns and to reduce water use, are required. Fundamentally, the focus will need to shift from more "crop per drop" to more "jobs per drop". Service and industrial sectors consume far less water per employment generated, supporting greater growth.

The participatory approach to prepare and apply an integrated river basin model as presented in this paper has potential to support improved strategic planning for the Ganga Basin as shown by the results for the scenario and strategies presented. A similar approach can also have added value to support strategic planning for other large river basins in South Asia and the rest of the world. The components of the integrated model should then be modified to reflect the river basin, the issues and the possible interventions.

Author Contributions: Conceptualization, M.v.d.V., K.C.A.B. and W.Y.; Methodology, M.v.d.V., P.B., K.C.A.B., M.H., G.H., M.v.O., B.O., F.R. and A.W.; Project administration, K.C.A.B.; Software, M.v.d.V., P.B., M.H., G.H. and F.R.; Supervision, M.v.d.V., K.C.A.B. and W.Y.; Validation, M.v.d.V., P.B., M.H., M.v.O., F.R., R.N.S. and S.K.S.; Visualization, G.H.; Writing—original draft, M.v.d.V., P.B., K.C.A.B., M.H., G.H., M.v.O., B.O., F.R. and A.W.; Writing—review and editing, R.N.S, S.K.S. and W.Y.

Funding: This research was funded by South Asia Water Initiative under World Bank contract number 8005347. The funding support from the governments of Australia, Norway and the United Kingdom are gratefully acknowledged.

Acknowledgments: The authors thank the Central Water Commission, the Central Ground Water Board, the Central Pollution Control Board and the Indian Meteorological Department for making available their data. Furthermore, the authors are grateful to all institutions and individuals that have participated in the stakeholder process at national and state level. The Ministry of Water Resources, River Development and Ganga Rejuvenation is thanked for providing guidance to the implementation of the project.

Conflicts of Interest: The authors declare no conflicts of interest.

References

1. FAO Aquastat. Available online: http://www.fao.org/land-water/databases-and-software/aquastat/en/ (accessed on 14 June 2019).
2. Office of the Registrar General & Census Commissioner, India. Available online: http://censusindia.gov.in/ (accessed on 14 June 2019).
3. Sinha, R. Ecology of the River Ganga -Issues and Challenges. Proceedings of the National Seminar on Impact of Technology on Society: Issues and Challenges, Patna, India; Available online: https://www.researchgate.net/profile/Ravindra_Sinha2 (accessed on 14 June 2019).
4. Namami Gange Programme. Available online: http://nmcg.nic.in/NamamiGanga.aspx (accessed on 14 June 2019).
5. Van der Vat, M.P. (Ed.) *Ganga River Basin Model and Information System, Report and Documentation*; Deltares: Delft, The Netherlands, 2018. Available online: http://cwc.gov.in/sites/default/files/ganga-river-basin-model-and-wis-report-and-documentation.pdf (accessed on 14 June 2019).
6. Ottow, B.; Bons, C.A.; Negi, R.S. *Final Stakeholder Engagement Report*; Deltares, AEcom and FutureWater Delft: Delft, The Netherlands, 2017.
7. Bons, C.A. (Ed.) Ganga River Basin Planning Assessment Report. In *Main Volume and Appendices*; Report 1220123-002-ZWS-0003; Deltares with AECOM: Delft, The Netherlands; FutureWater for the World Bank: Wageningen, The Netherlands; The Government of India: New Delhi, India, 2019. Available online: http://cwc.gov.in/sites/default/files/ganga-river-basin-assessment-report.pdf (accessed on 14 June 2019).
8. Basco-Carrera, L.; Warren, A.; van Beek, E.; Jonoski, A.; Giardino, A. Collaborative modelling or participatory modelling? A framework for water resources management. *Environ. Model. Softw.* **2017**, *91*, 95–110.
9. Vennix, J.A.M. Group model-building: Tackling messy problems. *Syst. Dyn. Rev.* **1999**, *15*, 379–401. [CrossRef]
10. Hare, M. *A Guide to Group Model Building: How to Help Stakeholders Participate in Building and Discussing Models in Order to Improve Understanding of Resource Management*; Seecon Report Seecon03/2003; Seecon Deutschland GmbH: Osnabrück, Germany, 2003.
11. Creighton, J.L. *How to Conduct a Shared Vision Planning Process*; Report 10-R-6; U.S. Army Corps of Engineers Institute for Water Resources: Washington, DC, USA, 2010.
12. INRM Consultants Pvt. Ltd. *Water Systems Modelling for Ganga Basin*; Final Report, Contract No. 7152490; INRM Consultants Pvt. Ltd.: New Delhi, India, 2010.
13. Institute of Water Modelling. *Ganges River Basin Modelling*; Final Report; Institute of Water Modelling: Dhaka, Bangladesh, 2010.
14. Indian Institutes of Technology. Surface and Groundwater Modelling of the Ganga River Basin. In *GRBMP: Ganga River Basin Management Plan*; Indian Institutes of Technology: Delhi, India, 2014.
15. Choudhury, S. Implementation the Key to Ganga Cleaning. *The Statesman*. 18 July 2017. Available online: https://www.thestatesman.com/opinion/implementation-the-key-to-ganga-cleaning-1500411351.html (accessed on 14 June 2019).

16. Terink, W.; Lutz, A.F.; Simons, G.W.H.; Immerzeel, W.W.; Droogers, P. SPHY v2.0: Spatial Processes in Hydrology. *Geosci. Model Dev.* **2015**, *8*, 2009–2034. [CrossRef]
17. Terink, W.; Lutz, A.F.; Immerzeel, W.W. *SPHY V2.0: Spatial Processes in Hydrology: Model Theory and SPHY Interface (v1.0) Manual*; FutureWater: Wageningen, The Netherlands, 2015.
18. Wflow. Available online: https://oss.deltares.nl/web/wflow/home/-/blogs/welcome-to-the-wflow-webpage (accessed on 14 June 2019).
19. Lutz, A.F.; Immerzeel, W.W.; Shrestha, A.A.; Bierkens, M.F.P. Consistent increase in High Asia's runoff due to increasing glacier melt and precipitation. *Nat. Clim. Chang.* **2014**, *4*, 587–592. [CrossRef]
20. Lehner, B.; Verdin, K.; Jarvis, A. *HydroSHEDS Technical Documentation Version 1.0*; World Wildlife Fund US: Washington, DC, USA, 2006.
21. OpenStreetMap. Available online: http://openstreatmap.org (accessed on 14 June 2019).
22. Bhuvan Indian Geo-Platform if ISRO. Available online: https://bhuvan-app1.rsc.gov.in (accessed on 14 June 2019).
23. Defourny, P.; Vancutsem, C.; Bicheron, P.; Brockmann, C.; Nino, F.; Schouten, L.; Leroy, M. GLOBCOVER: A 300 m global land cover product for 2005 using ENVISAT MERIS time series. In Proceedings of the ISPRS Commission VII Mid-Term Symposium: Remote Sensing: From Pixels to Processes, Enschede, The Netherland, 8–11 May 2006; pp. 8–11.
24. FAO Digital Soil Map of the World. Available online: http://www.fao.org/land-water/land/land-governance/land-resources-planning-toolbox/category/details/en/c/1026564/ (accessed on 14 June 2019).
25. De Boer, F. *HiHydroSoil: A High Resolution Soil Map of Hydraulic Properties*; FutureWater: Wageningen, The Netherlands, 2015.
26. Arendt, A.; Bliss, A.; Bolch, T.; Cogley, J.G.; Gardner, A.S.; Hagen, J.O.; Hock, R.; Huss, M.; Kaser, G.; Kienholz, C.; et al. *Randolph Glacier Inventory [5.0]: A Dataset of Global Glacier Outlines, Version 5.0*; Global Land Ice Measurements from Space (GLIMS): Boulder, CO, USA, 2015.
27. India Meteorological Department. Available online: http://www.imd/gov.in (accessed on 14 June 2019).
28. Weedon, G.P.; Balsamo, G.; Bellouin, N.; Gomes, S.; Best, M.J.; Viterbo, P. The WFDEI meteorological forcing data set: WATCH Forcing Data methodology applied to ERA-Interim reanalysis data. *Water Resour. Res.* **2014**, *9*, 7505–7514. [CrossRef]
29. Weedon, G.P.; Gomes, S.; Viterbo, P.; Shuttleworth, W.J.; Blyth, E.; Österle, H.; Adam, J.C.; Bellouin, N.; Boucher, O.; Best, M. Creation of the WATCH Forcing Data and Its Use to Assess Global and Regional Reference Crop Evaporation over Land during the Twentieth Century. *J. Hydrometeorol.* **2011**, *12*, 823–848. [CrossRef]
30. Srivastava, A.K.; Rajeevan, M.; Kshirsagar, S.R. Development of a high resolution daily gridded temperature data set (1969–2005) for the Indian region. *Atmos. Sci. Lett.* **2009**, *10*, 249–254. [CrossRef]
31. Van der Krogt, W.N.M. *RIBASIM Version 7.00, Technical Reference Manual*; Deltares Report T2478.00; Deltare: Delft, The Netherlands, 2008; p. 158.
32. Van der Krogt, W.N.M.; Boccalon, A. *River Basin Simulation Model RIBASIM Version 7.00, User Manual*; Deltares Project 1000408-002; Deltare: Delft, The Netherlands, 2013; p. 213.
33. CWC and NRSC. *Ganga Basin, Version 2.0*; Government of India, Ministry of Water Resources: New Delhi, India; Central Water Commission (CWC): Bhubaneswar, India; National Remote Sensing Centre (NRSC): Telangana, India, 2014; p. 259.
34. India WRIS. Available online: http://www.india-wris.nrsc.gov.in/wris.html (accessed on 14 June 2019).
35. Land Use Statistics Information System of the Ministry of Agriculture and Farmers Welfare. Available online: http://aps.dac.gov.in/LUS/Public/Reports.aspx (accessed on 14 June 2019).
36. Crop Science Division of the Indian Council of Agricultural Research. Available online: http://eands.dacnet.nic.in/At_A_Glance-2011/Annex-IV.xls (accessed on 14 June 2019).
37. Gupta, S.K.; Deshpande, R.D. Water for India in 2050: First-order assessment of available options. *Curr. Sci.* **2004**, *86*, 1216–1224.
38. Nag, A.; Adamala, S.; Raghuwanshi, N.S.; Singh, R.; Bandyopadhyay, A. Estimation and Ranking of Reference Evapotranspiration for Different Spatial Scales in India. *J. Indian Water Resour. Soc.* **2014**, *34*, 35.
39. Hajare, H.V.; RAMAN, D.; DHARKAR, E.J. A new technique for evaluation of crop coefficients: A case study. In Proceedings of the 2nd IASME/WSEAS International Conference on Water Resources, Hydraulics & Hydrology, Portoroz, Slovenia, 15–17 May 2007.

40. Shankar, V.; Ojha, C.S.P.; Prasad, K.S.H. Irrigation Scheduling for Maize and Indian-mustard based on Daily Crop Water Requirement in a Semi-Arid Region. *Int. J. Biol. Biomol. Agric. Food Biotechnol. Eng.* **2012**, *6*, 476–485.
41. FAO Crop Water Information. Available online: http://www.fao.org/land-water/databases-and-software/crop-information/en/ (accessed on 14 June 2019).
42. Office of the Registrar General & Census Commissioner. *Population Projections for India and States 2001–2026*; Report of the Technical Group on Population Projections; The National Commission on Population: New Delhi, India, 2006.
43. Narain, S.; Srinivasan, R.K. *Excreta Matters, Volume 2, State of India's Environment*; A Citizen's Report; Centre for Science and Environment: New Delhi, India, 2012; Volume 7, p. 496.
44. CPCB. *Pollution Assessment: River Ganga*; Central Pollution Control Board, Ministry of Environment and Forests, Government of India: New Delhi, India, 2013.
45. CGWB. *Dynamic Ground Water Resources of India (As on 31st March 2011)*; Central Ground Water Board, Ministry of Water Resources, River Development & Ganga Rejuvenation, Government of India: Faridabad, India, 2014; p. 299.
46. Postma, L.; Boderie, P.; van Gils, J.; van Beek, J. Component Software System for Surface Water Simulations. *Lect. Notes Comput. Sci.* **2003**, *2657*, 649–658. [CrossRef]
47. DWAQ Open Source Numerical Water Quality Model. Available online: https://oss.deltares.nl/web/delft3d/delwaq (accessed on 14 June 2019).
48. Smits, J.G.; van Beek, J. A Generic Eutrophication Model Including Comprehensive Sediment-Water Interaction. *PLoS ONE* **2013**, *8*, e68104. [CrossRef] [PubMed]
49. Metcalf & Eddy. Wastewater engineering: Treatment disposal reuse. In *McGraw-Hill Series in Water Resources and Environmental Engineering*; McGraw-Hill: New York, NY, USA, 1979; ISBN 007041677X.
50. CSE. *Excreta Matters A Profile of the Water and Sewage Situation in 71 Indian Cities A Survey by the Centre for Science and Environment*; Banerjee, S., Chaudhuri, J., Eds.; Centre for Science and Environment: New Delhi, India, 2012; ISBN 978-81-86906-56-9.
51. Central Public Health and Environmental Engineering Organisation (CPHEE). *Manual on Sewerage and Sewage Treatment Systems, Part B Operation and Maintenance*; Ministry of Urban Development Government of India: New Delhi, India, November 2013. Available online: http//moud.gov.in (accessed on 14 June 2019).
52. CPCB. *Management of Distillery Wastewater*; Resource Recycling Series: RERES/4/2001-2002; CPCB: New Delhi, India, 2001.
53. Qasim, W.; Mane, A.V. Characterization and treatment of selected food industrial effluents by coagulation and adsorption techniques. *Water Resour. Ind.* **2013**, *4*, 1–12. [CrossRef]
54. Irshad, A.; Talukder, S.; Selvakumar, K. *Current Practices and Emerging Trends in Abattoir Effluent Treatment in India: A Review*; Department of Livestock Products Technology (Meat Science), Veterinary College and Research Institute: Namakkal, India, 2015. [CrossRef]
55. Yadav, S.; Yadav, N. Physicochemical study of Paper Mill Effluent: To asses pollutant release to Environment Research article. *Int. J. Environ. Sci.* **2014**, *4*, 1053–1057. [CrossRef]
56. Central Pulp & Paper Research Institute. *Report on Water Conservation in Pulp & Paper Industry*; Cess Grant Authority (Development Council for Pulp, Paper & Allied Industries): Saharanpu, India, 2008.
57. Kumar Poddar, P.; Sahu, O. Quality and management of wastewater in sugar industry. *Appl Water Sci.* **2009**, *7*, 461–468. [CrossRef]
58. CPCB. *Environmental Standards for Ambient Air, Automobiles, Fuels, Industries and Noise Pollution Control Law Series: PCLS/4/2000*; CPCB: New Delhi, India, 2000.
59. Pathak, H.; Mohanty, S.; Jain, N.; Bhatia, A. Nitrogen, phosphorus, and potassium budgets in Indian agriculture. *Nutr. Cycl. Agroecosyst.* **2010**, *86*, 287–299. [CrossRef]
60. Central Public Health and Environmental Engineering Organisation (CPHEE). *Manual on Sewerage and Sewage Treatment Systems, Part A Engineering*; Ministry of Urban Development Government of India: New Delhi, India, November 2013. Available online: http//moud.gov.in (accessed on 14 June 2019).
61. Tare, V.; Bose, P. *Compendium of Sewage Treatment Technologies*; National River Conservation Directorate, Ministry of Environment and Forests, Government of India: New Delhi, India, 2009.

62. Vermeulen, M.P.T.; Roelofsen, F.J.; Minnema, B.; Burgering, L.M.T.; Verkaik, J.; Rakotonirina, A.D. *iMOD User Manual Version 4.3*; Deltares: Delft, The Netherlands, 2018; Available online: http://oss.deltares.nl/web/iMOD (accessed on 19 December 2018).
63. Harbaugh, A.W. *MODFLOW-2005, The U.S. Geological Survey Modular Ground-Water Model-the Ground-Water Flow Process*; Techniques and Methods 6-A16; U.S. Geological Survey: Reston, VA, USA, 2005.
64. Thenkabail, P.S.; Biradar, C.M.; Turral, H.; Noojipady, P.; Li, Y.J.; Vithanage, J.; Dheeravath, V.; Velpuri, M.; Schull, M.; Cai, X.L.; et al. *An Irrigated Area Map of the World (1999) Derived from Remote Sensing*; Research Report 105; International Water Management Institute: Colombo, Sri Lanka, 2006.
65. Delft-FEWS. Available online: http://oss.deltares.nl/web/delft-fews/ (accessed on 14 June 2019).
66. PostgreSQL/PostGIS. Available online: http://www.postgresql.org (accessed on 14 June 2019).
67. THREDDS. Available online: https://www.unidata.ucar.edu/software/thredds/current/tds (accessed on 14 June 2019).
68. NetCDF. Available online: https://www.unidata.ucar.edu/software/thredds/netcdf (accessed on 14 June 2019).
69. Ganga River Basin Planning. Available online: http://www.gangariverbasinplanning.com (accessed on 14 June 2019).
70. Delta Data Viewer. Available online: http://www.gangariverbasinplanning.com/ganga-information-system (accessed on 14 June 2019).
71. Richter, B.; Baumgartner, J.; Wigington, R.; Braun, D. How much water does a river need? *Freshw. Biol.* **1997**, *37*, 231–249. [CrossRef]

Article

Freshwater Ecosystems versus Hydropower Development: Environmental Assessments and Conservation Measures in the Transboundary Amur River Basin

Eugene A. Simonov [1,2,*], Oxana I. Nikitina [3,*] and Eugene G. Egidarev [3,4]

1 Rivers without Boundaries International Coalition, Dalian 116650, China
2 Daursky Biosphere Reserve, 674480 Nizhny Tsasuchey, Russia
3 World Wide Fund for Nature (WWF-Russia), 109240 Moscow, Russia
4 Pacific Geographical Institute of the Far Eastern Branch of the Russian Academy of Sciences, 690041 Vladivostok, Russia
* Correspondence: esimonovster@gmail.com (E.A.S.); onikitina@wwf.ru (O.I.N.)

Received: 18 June 2019; Accepted: 25 July 2019; Published: 29 July 2019

Abstract: Hydropower development causes a multitude of negative effects on freshwater ecosystems, and to prevent and minimize possible damage, environmental impact assessments must be conducted and optimal management scenarios designed. This paper examines the impacts of both existing and proposed hydropower development on the transboundary Amur River basin shared by Russia, China, and Mongolia, including the effectiveness of different tools and measures to minimize damage. It demonstrates that the application of various assessment and conservation tools at the proper time and in the proper sequence is the key factor in mitigating and minimizing the environmental impacts of dams. The tools considered include basin-wide assessments of hydropower impacts, the creation of protected areas on rivers threatened by dam construction, and environmental flows. The results of this work show how the initial avoidance and mitigation of hydropower impacts at early planning stages are more productive than the application of any measures during and after dam construction, that the assessment of hydropower impacts must be performed at a basin level rather than be limited to a project implementation site, and that the full spectrum of possible development scenarios should be considered. In addition, this project demonstrates that stakeholder analysis and robust public engagement are as crucial for the success of environmental assessments as scientific research is for the protection of river basins.

Keywords: hydropower; dam; damage; ecosystem; conservation measures; environmental assessment; environmental flows; GIS

---

1. Introduction

1.1. Freshwater Biodiversity and Dams

Although freshwater ecosystems cover less than 1% of the earth's surface, freshwater habitats are home to more than 10% of all known animals and about one-third of all known vertebrate species [1]. These ecosystems are also the most threatened: They are strongly affected by habitat modification, fragmentation and destruction, invasive species, overfishing, pollution, disease, climate change, etc. Freshwater ecosystem health is defined by its water quality and quantity [2], connectivity to other parts of the system and landscape, habitat condition [3], and diversity and abundance of plant and animal species [4]. Infrastructure development—especially dams—has caused a dramatic decline in the number of connected, free-flowing rivers [5]: Currently, there are more than 50,000 large dams worldwide [6]. Some of these dams are used for hydropower, which is the largest contributor to global renewable electricity generation, supplying 16.4% of the world's electricity from all sources [7]. During the process of hydropower project development, insufficient attention is paid to impacts on the environment (such as the disruption of the natural flow regime, fragmentation of the single river ecosystem, suppression of migration paths and the changing habitats of species, greenhouse gas emissions from reservoirs, changes in sediment flow and channel processes, changes in the microclimate, transformation of biological and chemical properties of the water body) and measures to minimize these [8].

1.2. Assessment Hierarchy and Sequencing for Hydropower

Hydropower projects should seek to minimize their impact on natural ecosystems and ecosystem services while optimizing the project's energy generation potential. Adherence to the mitigation hierarchy from the earliest planning phase and throughout the project life cycle can achieve more sustainable hydropower generation [9].

The evolving context of available energy alternatives is implicitly present in hydropower discourse. From the beginning of the 21st century, wide recognition of the urgent need to reduce greenhouse gas (GHG) emissions was the reason for proponents of large hydro and nuclear power to promote these technologies as the most promising climate-friendly options [10] that provided unique benefits outweighing negative impacts on ecosystems and local communities [11]. With the emergence of massive wind and solar production having the potential to outcompete more expensive technologies, the same groups now describe hydropower as an important enabler and stabilizer for the mass deployment of intermittent renewable technologies [12]. Other industry-led entities, such as the Global Energy Interconnection and Development Organization, dare to propose world-wide renewable energy systems, where a unified high-voltage grid supplies load centers from remote solar, wind, and hydropower "energy bases" [13]. Therefore, an analysis of alternatives at the energy system level is no longer a formality, but a necessary first step when considering impact assessments in the energy sector.

The strategic environmental assessment (SEA) is accepted worldwide as the instrument to facilitate a more comprehensive analysis of energy development, either standing alone or in combination with integrated planning approaches, such as integrated water resources management (IWRM) or integrated river basin planning [14]. According to the Netherlands Commission for Environmental Assessment (NCEA), about 30 SEAs have been executed supporting decision-making on hydropower development, with most of those completed in Asian countries [15].

In theory, three distinct phases can be identified in the assessment and decision-making that leads to large dam development:

1. Strategic planning to explore ways to fulfill societal needs with dams considered as one of the options;
2. Dam prefeasibility studies, including siting studies;
3. Dam feasibility and design after the selection of the preferred dam option [16].

In practice, only phase three is subject to an environmental assessment with formalized public participation, and hydropower project proponents will usually perform the previous phases without public consultations.

The objective is to compare environmental risks that have been mitigated at different stages of hydropower development and analyze the effectiveness of conservation measures. The hypothesis is that initial avoidance and mitigation of hydropower impacts undertaken at early planning stages are more effective in reducing damage to ecosystems in a given basin than measures undertaken during and after a dam is constructed.

The effectiveness of application of specific assessment methodologies can be judged only in the institutional and social context of its use. Governance systems as such may require formalized assessment before decisions to invest in hydropower project are made. According to the Netherlands

Commission for Environmental Assessment (NCEA), drivers (or root causes) for negative effects of dam development decisions are mostly related to deficiencies in the wider governance context such as neglect of strategic and system-based studies, favoring large dams above potentially more sustainable options without justification, and national public governance system incapable of correcting this situation due to late or insufficient involvement in the development process [16].

1.3. Importance of Stakeholder Engagement

Stakeholder analysis and proper engagement are crucial for environmental impact assessments (EIAs) because outcomes depend on the timely participation of stakeholders in the process, which can increase the willingness of the project developers to implement recommendations resulting from assessments. The social and cultural dimensions of dam development are widely documented and recognized [17] but are still rarely considered in assessments, especially at early stages of project identification [18]. The most neglected dimensions are an analysis of stakeholders involved in and affected by hydropower construction, and the development of tools to promote early stakeholder involvement into decision-making. In emerging economies, strategic planning is often expected to be based on objective science [19], rather than on the demands of various stakeholders [20].

While negative impacts on local communities partly resulting from incomplete assessments are well described in the literature [21], analyses of response strategies available to various stakeholder groups excluded from the assessment process are limited. One research work compared the more robust indigenous participatory mechanisms in Canada to those in Russia, where the development of the large Evenkiiskaya dam was halted partly due to active resistance from the indigenous community that was denied access to the official decision-making process [22]. Similar research on the Mekong [23] and rivers of Myanmar [24] has shown that the various facets in civil society that are denied access to decision-making may engage in developing their own disruptive strategies, including alternative assessment frameworks, and this may lead to drastic adjustments to project development processes driven by project proponents. Even companies strictly following national guidelines (and even best international practices) for public consultations during an EIA process still often experience fierce opposition from local communities, mainly when communities believe that the consultation process is being used as a tool to force hydropower projects in areas where communities do not agree in principle with dam building, or they see dams as a tool used by outside forces (e.g., central government) to impose control over territory/resources that local communities want to manage themselves [25]. Therefore, caution regarding public participation among project developers is quite understandable and underlies the strategy to make investment plans public only at the last stages of decision-making. However, a study that modeled such a process in Nepal still concluded that conflicts with local populations (inevitably embedded in dam projects) should be revealed and addressed at early stages of development and that by doing so, developers will save money and decrease risks [26]. The authors emphasized the lower implementation costs and higher chance for success when plans are changed at an early stage in the project cycle.

Many local stakeholders potentially affected by hydropower development are also unprepared to devote their resources to assess strategic planning documents. The majority of strategic planning documents, such as basin management plans, are not viewed by the public in developing/emerging economies as the real foundation for future decision-making, which is subject to the discretion of various government officials. Local actors most readily address problems that have already materialized and threatened their livelihoods, such as proposals to build a specific dam.

1.4. Amur River Basin: Biodiversity and Hydropower

An evaluation of the impacts of hydropower dams in the transboundary Amur basin is used to support the hypothesis that initial avoidance and mitigation of hydropower impacts can lead to better conservation outcomes. Amur is the largest transboundary river system in Northeast Asia, flowing through Mongolia, China, and Russia and forming a natural border between China and Russia (see

Figure 1). The river basin is famous for rare waterfowl, big cats, and endemic fish. The basin is still an arena for massive fish migration along the main stem and tributaries, with salmonids, sturgeon, and lamprey being three important examples [27]. The floodplains of the Amur and its tributaries create a belt of wetlands with high biodiversity value. The basin area is included in the list of Global 200 Ecoregions of the World, which are a priority for conservation efforts [28].

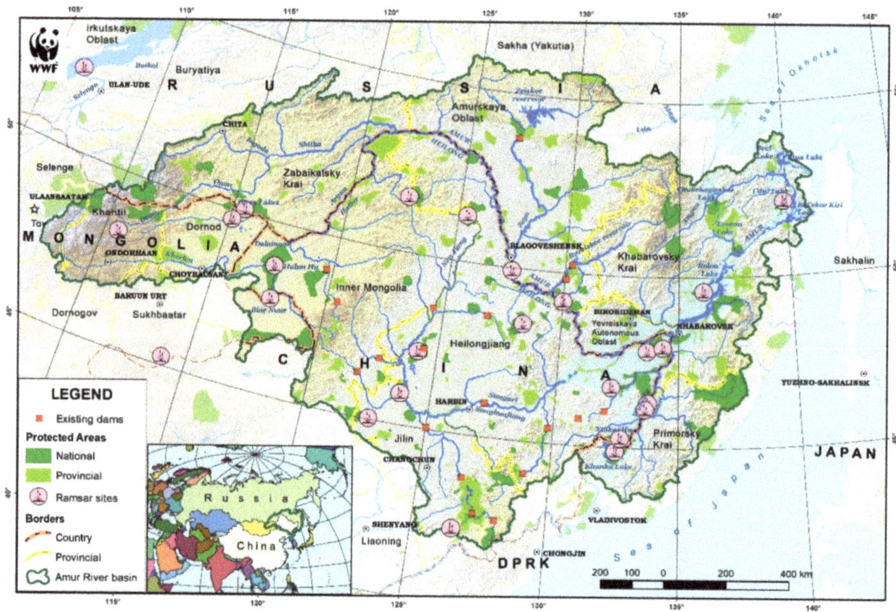

Figure 1. The Amur River basin and its protected areas. © WWF-Russia.

In 2019, there were approximately 100 dams in the basin, including 19 large dams, of which 3 are located in the Russian part on the Zeya and Bureya rivers. The total generation is almost 22 GWh per year, with the three large Russian dams producing 13.7 GWh per year.

In China, all the most promising sites suitable for large hydropower had already been used by the beginning of the 21st century. Recently built medium-sized dams (e.g., Hadashan on the Second Songhua River) have pursued multiple purposes, with hydropower being subordinate to irrigation, water supply, navigation, etc. Since 2015, Chinese authorities have imposed restrictions on new small hydropower construction and have started the assessment, reconstruction, and removal of smaller dams in many river basins, including the headwaters of the Second Songhua River in the Amur basin [29].

In Russia, the Amur basin was the first region where intensive hydropower construction was resumed after the economic crisis of the 1990s: The Bureya hydropower dam was put into operation in 2003 to complement the Zeya dam built in 1975, and the Lower Bureya dam started generation in 2017. The flow regime of the Zeya and Bureya have changed significantly, which has resulted in the alteration of the natural floodplain ecosystems on both rivers. This has caused a decline in typical floodplain communities [30], habitats of cranes and storks [31], and refuges for fish species [32]. Dams have become a barrier to migration, and chum salmon (keta), lamprey, and whitefish have disappeared above the dams [33]. The continuing degradation processes of the floodplain system of the Amur under the cumulative influence of the Zeya and Bureya hydropower dams has been further exacerbated below the mouth of the Songhua River, where the flow regime has undergone additional anthropogenic changes due to the construction of hydro-engineering structures in China [34].

1.5. Initial Dialogue between CSOs and Energy Industry

The hypothesis is that the basin-wide assessment of hydropower impacts is an effective tool in early planning when analyzing environmental costs of hydropower development and comparing possible development scenarios.

In Russia, the official environmental impact assessment methods used for hydropower projects do not allow for the analysis of complex impacts from dam construction on the ecosystem of the basin as a whole and do not compare different scenarios to optimize development [35]. For comparison, in 2011, Chinese regulations were issued for basin-wide hydropower schemes that require a basin-wide strategic assessment to precede individual dam EIAs [36]. However, the application of such assessments with meaningful public participation is yet to be fully implemented.

The lack of legal requirements and established practices for comprehensive basin-wide assessments of hydropower impacts in early 21st century Russia has become a serious obstacle to productive dialogue between the newly created state-owned Hydro-OGK hydropower company, in charge of 70% of hydropower dams (later renamed RusHydro Co.), and a wide coalition of non-governmental organizations (International Socio-Ecological Union, Greenpeace, WWF-Russia, Russian Bird Conservation Union, and other NGOs.). In the aftermath of the World Commission on Dams Report [17], the two sides sought to agree on safeguards to prevent negative impacts from new hydropower projects under specific conditions for Russia. NGO and industry experts compiled a "comprehensive list of impacts from hydropower dams and reservoirs on the environment", which identified 9 groups of problems and more than 100 specific potential impacts. An analysis of causal relationships between a multitude of effects and issues demonstrated that the majority of these were determined or strongly correlated with three fundamental environmental changes brought about by large reservoir development. These changes included (1) freshwater ecosystem fragmentation/blocking of the natural movement of biological organisms, sediments, nutrients, etc.; (2) the augmentation of natural flow dynamics of matter and energy in the river network; and (3) the creation of vast lake-like habitats serving various human activities that replaced natural river valleys. Discussions between NGOs and hydropower corporations highlighted mutual interest in identifying and ranking hydropower development options in each large river basin according to the potential severity of local and basin-wide impacts on ecosystems. This could assist both the industry and NGO community in defining long-term priorities for action and predicting and preventing possible conflicts and negative consequences. Therefore, it was suggested that the optimization of hydropower development in a large river basin first requires an assessment of impacts resulting from these three fundamental changes under different hydropower development scenarios. Such an assessment scheme was later designed by NGO experts and was initially welcomed by RusHydro management, although it was not explicitly used by the company [37].

2. Materials and Methods

The dialogue between RusHydro and leading civil society organizations (CSOs) held in 2007 helped to identify and introduce into the national policy debate key tools for preventing hydropower's negative impacts on nature and people who became the foundation for CSO activities in this field for the following decade. In this paper, we explore how some of those tools were applied to the Amur basin.

The tools to prevent adverse dam impacts employed in the Amur basin by civil society included the following:

1. A strategic basin-wide assessment of hydropower impact;
2. The creation of protected areas on the rivers that were threatened by hydropower development;
3. Environmental flows.

The methodology underlying each of the tools designed for trial use in the Amur River basin is described in this section.

2.1. Strategic Basin-Wide Rapid Assessment of Hydropower Impacts

During preliminary phases of hydropower development planning, it is important to determine the scale of impacts from each potential dam and rank proposed projects and multidam development scenarios according to the degree of their environmental impacts. A basin-wide assessment of hydropower options allows for a comparison of dam sites in terms of potential effects on connectivity and downstream flow regimes. Somewhat similar basin-wide optimization approaches proposed by The Nature Conservancy (TNC) [38] and other research papers on the global footprint of dams [39] inspired the formulation of specific minimum requirements for the rapid assessment of basin-wide environmental impacts of existing and planned hydropower dams. Such an approach could yield the most significant gains if used to guide development in large basins not yet significantly altered by hydropower, and therefore the transboundary Amur basin was considered the priority among basins of Russia.

Rapid assessment methodology is focused on simple modeling of potential impacts from dams on river ecosystems and is intended to guide developers on how to choose options with the least negative environmental effects. When assessing the cumulative impacts of several dams on the ecological condition of the basin, first and foremost, we considered the following broad impact factors, which have been jointly identified as the most important by NGOs and industry experts:

1. The alteration of flow regimes and ecosystems downstream of dams that affects the three-dimensional interaction of the river and valley;
2. The transformation of riverine habitats in the region through their replacement by water reservoirs;
3. The fragmentation of river networks, including the disruption of migration routes of species and material transport.

As demonstrated by previous assessments of river basins around the world, these three factors are associated with the majority of observed and predicted consequences of dam building for aquatic ecosystems, e.g., mapping the world's free-flowing rivers [5], global threats to human water security and river biodiversity [40], and restoring environmental flows by modifying dam operations [41].

To assess and compare the impacts of multiple scenarios on the river basin, three pairs of proxy indicators for impact were designed, so that each of the main factors could be expressed both in absolute and relative values.

2.1.1. Flow Regime Alteration and Floodplain Transformation Downstream from Dams

The most important objective should be to protect the Amur River and its floodplains, which contain the main biological resources and ecological services as well as the natural support base for the local communities of the Southern Far East. The methodology for evaluating the socioecological impacts and criteria for future dam construction should be developed based on the main concern [42]. The degree of floodplain ecosystem transformation downstream from dams reflects the consequences of altering the natural hydrological regime of a river. The index is based on the relationship between the storage volume (live volume) of a reservoir(s) and the total annual river flow volume at the dam's location, which is a consensual index used in many studies. However, for the Amur River, it is expressed not just as the percentage of the flow volume that can be withheld in the dam's reservoirs upstream of a given river's reach (index commonly known as the degree of regulation (DOR), see Reference [5]), but is further multiplied by the area of the floodplain ecosystem at a given reach. The resulting index, *Imp_fl* (km^2), is proportional to both the area of vulnerable habitats and the degree of impact in a given river reach (stretch). To characterize the basin-wide situation, it is summed across all river reaches of the basin/sub-basin (*ALT_fl* (km^2)). When related to the original unaltered floodplain area of a given river reach, it approximates the share of affected floodplains as a percent: *Imp_fl* (%). For each scenario, it is summed across all river reaches of the basin/sub-basin. To characterize the basin-wide alteration of floodplains, the ratio between *ALT_fl* (km^2) and total area of natural floodplains ($\sum Sfl$) in the given basin/sub-basin is used and is expressed as a percent: $100 \times \sum Imp\_fl/(\sum Sfl) = ALT\_mean$ (%).

The degree of floodplain ecosystem transformation provides an index to measure how strongly a dam or set of dams can affect the natural floodplains and their ecosystem services.

Floodplains are singled out as the most ecologically important habitat affected by flow regulation, which is valid for the Amur basin, but applying the assessment methodology in a different river ecosystem may require focusing on different key habitat types. The biota and ecosystems of the rivers in the Amur catchment are strongly dependent on the floods that are regulated by dams.

2.1.2. Transformation of Riverine Habitats by Reservoirs

Any reservoir is an anthropogenic feature replacing the essential socioecological landscapes: river valleys. It is assumed that the larger the surface of the water reservoirs and the greater their share in all water surfaces of the river system, then the more they transform aquatic and terrestrial ecosystems. The surface of the existing river system and the modeled surface of planned reservoirs as well as the surface of all freshwater ecosystems upstream from the dam preceding reservoir formation (during the low-flow period) is calculated. Reservoir surface areas are expressed as *Imp_res* (km^2). Transformation by the reservoir is calculated as the percent ratio between reservoir surface area and surface area of all aquatic ecosystems of a given sub-basin before reservoir formation *IMP_reservoir* (%).

The value of this measure, when applied to the Bureya and Zeya dams, can be seen on the impacts of fish species and stocks. The Bureya dam and Zeya dam reservoirs together occupy 3160 km^2, which is roughly equal to 45% of the total water surface in the Middle Amur Freshwater Ecoregion in Russia. In China, all 12 hydropower reservoirs of the basin occupy only half of that area. The Zeya and Bureya reservoirs have low-quality water, in part due to inundation by massive volumes of vegetation, soil, and peat. Before the Zeya dam construction, the composition of the fish fauna of the Upper Zeya basin in 1970 included 38 species, but by 2007 the fish fauna of the Zeya reservoir was reduced to 26 species [43]. Fish stocks of the Zeya reservoir have been seriously depressed for many years.

2.1.3. Blocked Sub-Basins: A Measure of Basin Fragmentation

River fragmentation indices typically measure the degree to which river networks are fragmented longitudinally by infrastructure, such as hydropower dams. Fragmentation prevents effective ecological processes that depend on longitudinal river connectivity, including the transport of organic and inorganic matter and upstream and downstream movements of aquatic and riparian species [5,39].

A simple measure of the fragmentation of the river basin is a percentage of the basin area that is cut off from the sea by dams. This study takes as a proxy measure the area (*Imp_bl* (km^2)) or share (*IMP_block* (%)) of the basin disconnected by dams from the sea. In other more sophisticated versions of this methodology (not presented in this paper), additional fragmentation indices have also been tested that measure the degree of partition in many disconnected sub-basins [44]. For purposes of this rapid assessment, those additional indices do not reveal new trends but reinforce the emphasis on fragmentation. In a more nuanced study with a detailed comparison of multidam scenarios, the use of such indices would be more justified.

A blocked basin index was justified as a proxy index in our analysis because the global value of the Amur Freshwater Ecoregion was primarily attributed to the abundance of diadromous fish species that suffer from disruption of migratory routes to the sea [27]. For example, upstream from the Zeya and Bureya dams, the Amur sturgeon, kaluga, keta salmon, Japanese lamprey, and other migratory species have already disappeared. Taken together, by 2011, the Zeya and Bureya dams blocked 9% of the Amur catchment area, while all of the existing dams in China blocked an additional 22–23%. This means that nearly one-third of the Amur River system has already been isolated from the sea and can no longer sustain key migratory species, e.g., diadromous fish.

For calculations, the analyzed river system was divided into reaches (stretches) limited by existing and potential dams and confluences with tributaries. Since the analysis was applied only to large dams/reservoirs, the inquiry was limited to streams having a catchment area of more than 10,000 km^2 (with 3–4 exceptions where large dams were planned in catchments of a smaller size).

Each stretch with a complex of natural characteristics was defined as a key basic unit in the multifactor analysis. The whole Amur basin was delineated into 214 units/stretches (see Figure S1), to which we attributed characteristics of the respective elementary sub-basins. To define stream direction, borders of watersheds, the size of planned reservoirs, hydrographic models, and data from the Shuttle Radar Topographic Mission (SRTM) [45] were used.

Satellite imagery was processed using Erdas Image software, while expert interpretation of water bodies and floodplains was performed using ArcGIS 10.5. The sources of cartographic information were Landsat Aster and Sentinel-2 satellite imagery, SRTM water body data (SWBD) [46], and vector topographic maps (1:500,000–1:100,000).

Characteristics of hydropower engineering projects have been derived from planning documents, river basin development schemes, technical documentation for dam design, etc. Annual discharge at each reach was derived from official Soviet hydrological bulletins and literature on hydropower projects and plans, e.g., hydropower master plans [47], natural resource allocation studies [48], bilateral water management schemes [49,50], and five-year plans [51]. The remaining data gaps were filled by modeling/approximation.

From all available sources, a list of 84 possible large dam locations in the Amur River basin was compiled, from which 45 projects had data sufficient to perform a basin-wide impact analysis (see Figure 2).

The characteristics were attributed to each river reach (stretch) and related sub-basin as appropriate. The three proxy indicators of impact were calculated at each river stretch and/or each sub-basin for each assessed basin-wide hydropower development scenario (see Table S1).

The values of the three indicators at the river mouth (or at the terminal point of any sub-basin/catchment) were taken as a proxy for a basin-wide score for a given scenario in this catchment. To characterize the degree of alteration from one value (following the advice of Dr. Alexander Martynov), an integrated index was also developed:

$$INT3 = \sqrt[3]{(ALT\_mean) \times (IMP\_reservoir) \times (IMP\_block)}.$$

For comparisons between basin-wide scenarios, the three proxy indicators were used as well as the integrated index calculated at the Amur River estuary.

To relate electricity production to the potential impact, all proxy indices were divided by the annual electricity production expected in a given scenario. The impact per unit of energy generation is an important measure of efficiency in terms of environmental economics [52].

All existing and possible dams and hydropower development scenarios (combinations of dams described in planning documents) in the Amur River basin were ranked according to their potential environmental impacts and efficiency relative to energy production (see Table S2).

For the sustainable development of hydropower in a given river basin, the aim is to minimize both potential impacts and their relative measure per unit of energy production. This rapid assessment methodology enables a quick initial ranking of multiple hydropower development scenarios in a large river basin according to their key potential negative impacts. Therefore, it provides developers with an opportunity to concentrate further efforts on less risky development options, while conservationists can focus on the prevention of dam construction at locations where it may lead to the greatest negative impacts for natural freshwater ecosystems.

This is the first basin-wide assessment of hydropower impacts designed for the Amur basin and the first ever suggested in Russia. The main innovation of our methodology compared to earlier assessment methods designed for other basins has been to reduce the assessment to three key variables, which, once agreed upon, do not require subjective expert inputs in further scenario assessment. This enables a credible comparison of impacts from any number of hydropower development scenarios at a basin-wide scale, including existing and proposed dams. Despite simplification, the methodology still takes into account specific features of the basin in question (e.g., using the degree of floodplain ecosystem transformation index in place of the conventional degree of regulation index (DOR)). Our model has also been shown to have certain theoretical value by successfully relating various degrees of hydropower potential utilization to measurements of potential impacts associated with such development scenarios.

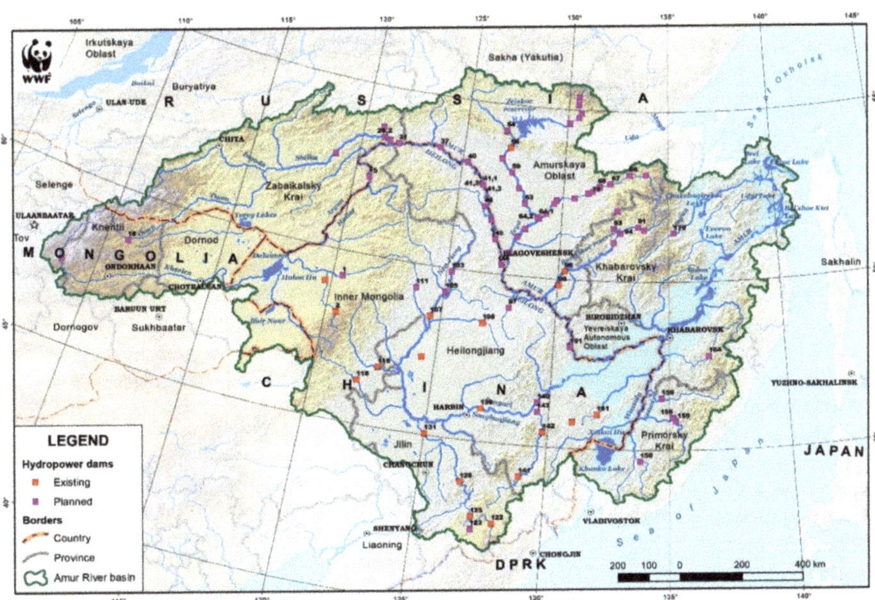

Figure 2. Locations of the existing and potential dams in the Amur River basin. © WWF-Russia.

2.2. Creation of Protected Areas at River Stretches Targeted for Hydropower Development

In the case of the Amur, the river stretches where hydropower construction is predicted to cause the most damage to the ecosystem of a given river basin are pre-emptively protected in perpetuity to constrain the future design of hydropower at such sites by legal regulations. Since sites known to be suitable for large dams are scarce and often located in river gorges with rich biodiversity, it is often possible to justify protected area development based on local natural values even without reference to its wider safeguarding function for a basin-wide river ecosystem. Since Russian legislation does not prescribe the design of specific biodiversity conservation measures as part of basin-wide water resources management planning, after the identification of critical sites, the rest of the work follows a standard methodology for the creation of protected areas.

In Russia, legally protected natural areas of different types can be established at a provincial and national level, while in China, four levels of government have such authority. In both countries, there are long-term plans for gazetting conservation areas, in which prospective protected areas should be included in order to be considered [53]. Field research by local scientists to collect data to justify the protection of a given site is usually required. In both countries, formal reports covering the scientific, economic, and management justification for a given protected area should be presented to a special commission formed by a relevant government agency. In addition, until 2019, Russia required the socioeconomic justification for protected areas (PA) establishment to be subject to an EIA procedure, which included public consultations with potentially affected stakeholders. The overall process of PA establishment could take anywhere from 2 to 10 years. However, once preservation on a natural hydrological regime and/or prohibition for massive infrastructure development is written into management regulations of a new protected area, it becomes a legal requirement. Removing or adjusting such prohibitions requires a new and lengthy procedure involving a new EIA and public consultations. Adjustments to a protection regime happen more often in China than in Russia, but still present serious legal and reputational challenges for proponents of dam building. In China, at least one precedent of the protected area being reconfigured to give way to dam building is known. However, this has not happened in Russia.

Protected areas have been created several times since the 1990s in the Amur basin to protect the most vulnerable sites. The Norsky Strict Nature Reserve in Amurskaya Province of Russia was the first such PA (successfully created in 1998) to safeguard the confluence of the Nora and Selemdzha rivers, where development of the Dagmarskaya dam was proposed in the early 1990s. The method was pioneered by a team of local conservation biologists led by Dr. Yury Darman [54].

This team of authors (to the best of our knowledge) is the first to apply this approach to protected areas for freshwater ecosystem conservation at the basin scale, namely to establish conservation areas at potential damming sites in a systematic manner.

2.3. Environmental Flows

Transformation of the flow regime downstream from the dam is among the most adverse impacts of dam construction on freshwater ecosystems. One of the compensatory measures to mitigate the impacts of dam construction is the implementation of environmental flow, which is the release of a specific volume of water from the reservoir to mimic the characteristics of the natural flow variability of water and sediments [7].

There are about 200 methods for determining environmental flows [55] being used globally. Regardless of the chosen method, it is important to characterize the flow regime and its intra- and interannual variability in natural conditions. These characteristics describe the water regime to which the ecosystem has been adapted.

In Russia, guidelines for the development of reservoir operating rules indicate that reservoirs should also be used for environmental flow releases [56]. The method prescribed in "Methodological Guidelines for Developing Standards to Assess Environmental Flow" is based on preserving the freshwater ecosystem in a state where its restoration potential is not disturbed by determining the critical flow volume [57]. Environmental flow is calculated based on the difference between the values of annual runoff and the volume of allowable water withdrawal. In this study, the applicability of various methods to freshwater ecosystems of the Amur basin was assessed, but we did not seek to design another specific methodology.

3. Results

The description of the results is characterized, to the extent possible, both from technical outcomes of the assessments as well as from the public participation and policy processes in which they were used.

3.1. Basin-Wide Hydropower Assessment

3.1.1. Main Assessment Findings

Using the basin-wide rapid assessment tool, how potential impacts depend on the degree and pattern of hydropower development was analyzed, with the following results.

At initial phases of development in unaffected basins, individual hydropower dams with similar production may have a more than 10-fold difference in potential basin-wide environmental impacts (Figure 3).

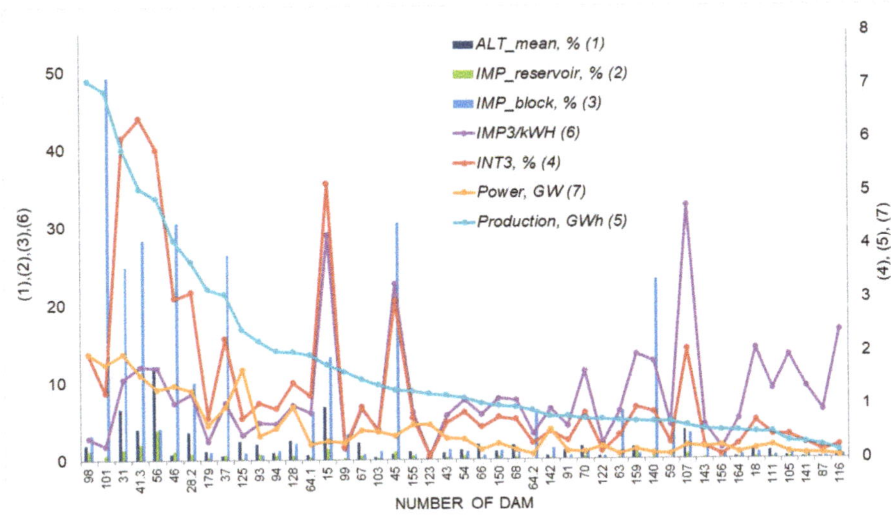

Figure 3. Impact of individual hydropower dams.

It was found that as the development of hydropower potential progresses, the difference between "best" and "worst" development options decreases. The scenarios selected for analysis in this paper are described in Table 1.

Table 1. Composition of the scenarios shown in Figures 4–6.

Figure 4		Figure 5		Figure 6	
Scenario 10% min	37, 38	Scenario Actual	2, 56, 98, 107, 108, 116, 118, 122, 125, 128, 139, 141, 142	Scenario actual	2, 56, 98, 107, 108, 116, 118, 122, 125, 128, 139, 141, 142
Scenario 10% max	41.2	Scenario Head 1	141, 159, 93, 155, 67	Scenario actual, add Bureya cascade	2, 56, 91, 93, 94, 98, 99, 107, 108, 116, 118, 122, 125, 128, 139, 141, 142
Scenario 25% actual	2, 56, 98, 107, 108, 116, 118, 122, 125, 128, 139, 141, 142	Scenario Head 2	18, 68, 125, 105, 54	Scenario actual, add low	2, 56, 64.1, 98, 99, 107, 108, 116, 118, 122, 125, 128, 139, 141, 142
Scenario 25% max	56, 98, 15, 28.2, 70, 107, 116, 159	Scenario Head 3	54, 70, 179	Scenario actual, add Shilka	2, 28.2, 56, 98, 107, 108, 116, 118, 122, 125, 128, 139, 141, 142
Scenario 25% min	56, 98, 91, 59, 94, 99, 93	Scenario Head 4	103, 150, 94,43, 142	Scenario actual, add Taipinggou	2, 56, 98, 101, 107, 108, 116, 118, 122, 125, 128, 139, 141, 142
Scenario 75% max	2, 15, 31, 37, 41.2, 45, 56, 98, 101, 107, 116, 118, 122, 125, 128, 139, 141, 142	Scenario Head 5	56	Scenario actual, add mainstream	2, 31, 41.3, 45, 56, 98, 101, 107, 108, 116, 118, 122, 125, 128, 139, 141, 142
Scenario 75% min	2, 15, 18, 43, 54, 56, 64.1, 66, 67, 68, 70, 87, 91, 93, 94, 98, 99, 103, 105, 107, 111, 116, 118, 123, 122, 125, 128, 139, 141, 142, 143, 150, 155, 156, 159, 164, 179	Scenario Down 1	41.3		
Scenario 100%	2, 15,18, 31, 37, 41.2, 43, 46, 54, 56, 59, 63, 64.1, 64.2, 66, 67, 68, 70, 87, 91, 93, 94, 98, 99, 101, 103, 105, 107, 111, 116, 118, 123, 122, 125, 128, 131, 139, 140, 141, 142, 143, 150, 155, 156, 159, 164, 179	Scenario Down 2	28.2, 45		

Figure 4 shows the scenarios with minimum and maximum impact, which were identified from all scenarios utilizing 10% of economic hydropower potential. In terms of the integral impact index, they differed six-fold, still showing ample potential for avoiding the most negative basin-wide impacts.

Figure 4 also shows an analysis of the actual development in the Amur basin (Scenario 25% actual), which was just below 25% of total economic hydropower potential, indicating that the same electricity production could have been achieved with a much smaller integral impact (Scenario 25% min), while the degree of impact in a scenario with maximum possible damage (Scenario 25% max) was 100% more than in Scenario 25% min, but only 10% more than in Scenario 25% actual.

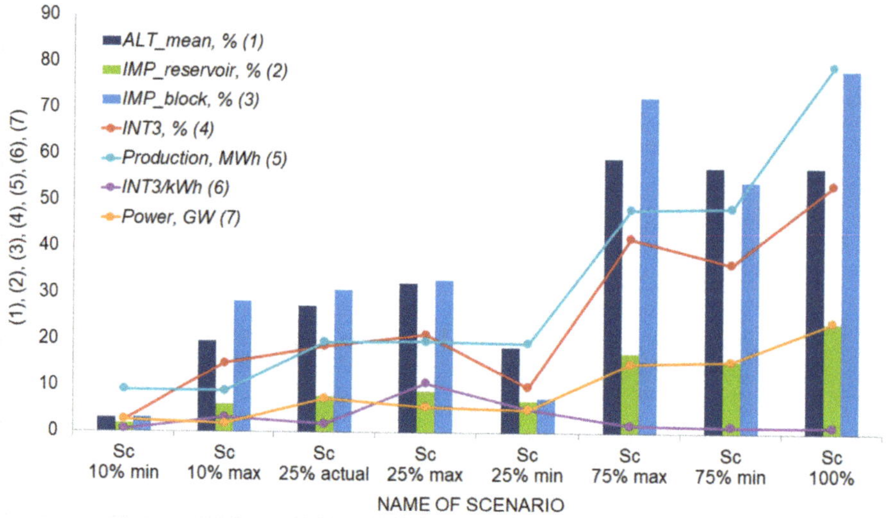

Figure 4. Scenario analysis and the corresponding impact due to different levels of development of the hydropower potential in the Amur basin.

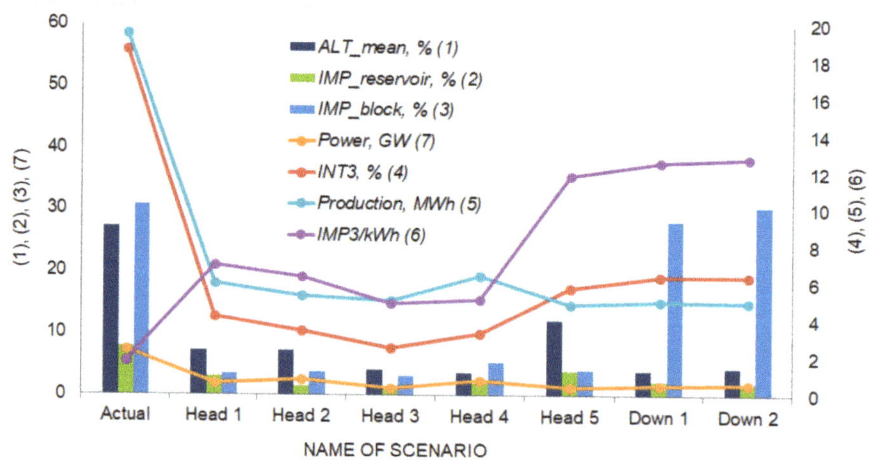

Figure 5. Relationship between the impact factors in scenarios when dams are located in the upper (Up) and lower (Down) parts of the Amur basin.

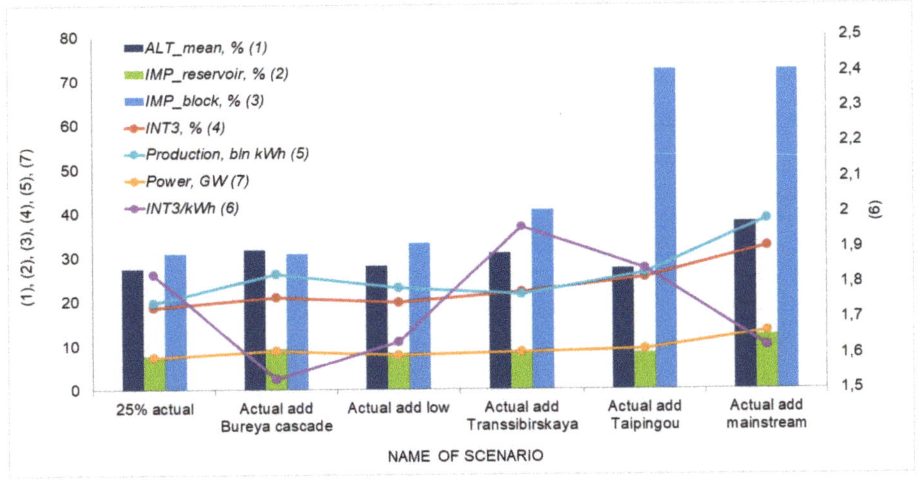

Figure 6. Potential scenarios of hydropower development based on the current situation.

Further modeling of scenarios with the development of 75% of technical potential resulting in maximum (Scenario 75% max) and minimal (Scenario 75% min) impact demonstrated that few choices remain between "worse" and the "worst" impacts, with only a 15% difference between the most damaging and least damaging options. Further development of "full economic potential" (Scenario 100%) led to only a 25% increase in integral impacts.

This shows that the common argument by hydropower proponents that "only 10% of hydropower potential is developed and therefore we need to develop much more" likely has an acute conflict with the notion of sustainable development because a greater proportion of basin-wide hydropower development leaves no room for "low-impact" sustainable development scenarios.

On the other hand, most impact has usually been achieved at relatively low levels of hydropower development (at least in the case of the Amur), while the additional development of dams in an already heavily affected basin brings significantly smaller additional incremental impacts. This also calls for the cautionary use of "efficiency" measures relating impact to the production of electricity, because it does not caution against excessive development in a given basin.

Dam location seems to be the leading factor defining the integral impact of hydropower development scenarios. According to this assessment methodology, "dams located downstream" are the dams that block a greater part of the overall basin. To test how this factor influences integral impact, dams were grouped in 8–9 scenarios, each having a total production of 5–6 GWh/year, with projects in a given scenario having similar "blocked basin" index values (see Figure 5). Some scenarios consisted of a single large dam on the main stem. Integral impact per unit production (*INT3*/kWh) increased for scenarios with dams located further downstream, primarily due to the increased share of the blocked basin (*IMP_block*, %), while the variation of the other two indices had no clear trend as we moved downstream. For "downstream" dams, the value of *INT3*/kWh was usually 2–2.5 times higher than for scenarios with "upstream" dams. If other aspects of fragmentation were included in the equation, dam location became an even more influential factor in defining the overall impact.

For the Amur basin, this assessment concludes that minimal additional impact is associated with the development of hydropower cascades on tributaries that already have dams. This minimizes fragmentation of the basin, interference with sediment flows, and other negative influences. Conversely, the development of dam cascades on the main stem of the Amur will lead to maximum negative impacts. The addition of even one relatively small dam (Khingansky—Taipinggou) on the main channel will lead to a sharp increase in the cumulative impacts of hydropower in the basin (see Figure 6).

However, "sustainable" development of hydropower in the Amur basin is significantly limited by the degree of cumulative impact from existing dams, which is already very close to the maximum possible impact that could be induced by any scenario with comparable electricity production.

3.1.2. Application of the Methodology in Business Interactions

In 2011, En+ Group/EuroSibEnergo Company, the largest private hydropower producer in Russia, announced plans to build a Transsibirskaya (Shilkinskaya) hydropower dam on the Shilka River. The company sought to diversify its business, which was primarily focused on aluminum production, and develop energy projects targeting the Chinese market and investors. The local population, as well as regional scientists and environmentalists, strongly opposed the proposed construction of a dam on the Shilka River, partly because this project had previously been proposed in the 1990s and its environmental impact assessment had shown significant negative impacts. WWF-Russia and other CSOs assisted local Russian stakeholders in developing a basin-wide campaign to communicate the potential effects and solicit feedback from civil society and local authorities in the five provinces of Russia along the Amur River [58]. To manage the conflict, the company decided to start a dialogue with civil society about sustainable hydropower options. In 2012, WWF-Russia and the En+ Group launched a research project to look into the potential for hydropower development in the Amur basin. The ultimate goal of the strategic assessment was to identify hydropower dam location options with the fewest environmental impacts for the whole Amur basin and the maximum social and economic benefit to the region. A list of 43 possible dam locations in the basin was determined, and potential environmental and socioeconomic impacts of the development on the region were analyzed.

The environmental part of the research assessed the condition of freshwater ecoregions, and one finding was that preservation of the unaltered Shilka ecoregion is essential for sustaining the ecological health of the Amur basin. The Transsibirskaya dam proposed on the Shilka River would increase the negative integral impact on the basin by 16% (see "Scenario actual, add Transsibirskaya" in Figure 6). In comparison, two additional dams on the Zeya and Bureya rivers would generate 20% more energy than the Transsibirskaya dam, but add only 4% to the integral impact on the Amur basin (see "Scenario actual, add low" in Figure 6). The assessment findings demonstrated that the increase in potential negative environmental impacts may differ by more than 10 times for development scenarios with the same additional electricity production. Dam location is the most decisive factor defining cumulative impacts on a basin scale. The best mitigation is to choose sites by using strategic assessments of the basin-wide plan and avoiding, at all costs, the development of sites that result in substantial basin-wide impacts. Existing high-impact dams severely limit the opportunities for further low-impact hydropower development and sustainable integrated water resources management.

The socioeconomic part attempted to assess the economic efficiency of the project: average prevented flood damage, macroeconomic budget efficiency, changes in employment, number of people resettled, changes in navigation conditions and turnover, losses to architectural heritage and archaeological sites, changes in fisheries, flooding, and economic flooding of objects. The socioeconomic part found that the existing Zeya and Bureya hydropower plants provide the biggest socioeconomic benefits compared to the other possible options. Among those possible dams, the Transsibirskaya dam along with the hydropower dam on the Lower Zeya and the Upper Bureya was listed as the next best option. However, the En+ Group stated that the joint assessment showed that the Transsibirskaya dam is an unbalanced/unsustainable option [44], and the company has not pursued this dam development any further.

According to the NCEA, this group is the first team of researchers and practitioners to document the process and outcomes of an SEA-like exercise carried out in real-life situations in partnership with a commercial company, without major involvement and mediation from state authorities. This makes the case study unique when compared to several other hydropower-related SEA cases that have been documented to date [11].

3.1.3. Interaction with Other Stakeholders and Further Policy Dialogue

The draft WWF&En+ Group basin-wide strategic environmental assessment report was subjected to a thorough review by local experts and representatives of various agencies. Their written comments and recommendations on the draft report were compiled and published with response matrices as an intrinsic part of the final assessment document. Besides the immediate value for the study and confirmation of its appropriateness and technical validity, such reviews played an important role in legitimizing, in the eyes of policy-makers, the results of an environmental assessment that had been initiated between an NGO and a private company and not sanctioned by authorities or academia. The company opted not to subject the economic part of the assessment to a similar review procedure, but published it with comments only from the members of the environmental assessment team.

At the same time, WWF-Russia and the En+ Group decided not to subject the report to public consultation with the local population because of the immense complexity of organizing it over the whole basin, which covers five provinces in Russia. Since the report was prepared in the Russian language, there was no formal involvement of stakeholders from China in its review either. The Rivers without Boundaries Coalition developed a summary of findings in English and shared it with relevant potential investors, including the China Yangtze Power and China Three Gorges corporations [59].

The same methodology of basin-wide hydropower assessment was applied to communicate with Russian and Chinese actors the potential impacts from a plan to develop 5–8 flood-control hydropower dams that had emerged in the aftermath of catastrophic flooding in the Amur basin in 2013. The CSOs argued that the proposed reservoirs had limited value for flow augmentation in comparison to existing natural floodplains [60], while the cumulative environmental impact from their development would be quite substantial [61]. By 2019, none of these flood control dam projects had progressed, and they are unlikely to re-emerge in the near future.

This was the first SEA-like methodology applied to basin-wide hydropower planning in Russia that received societal acceptance from various stakeholder groups, thus creating an important basis for future use of SEAs in water resources management in the country. To the best of the team's knowledge, this was also the first assessment focused on transboundary basins shared by China, and (although they did not actively participate) Chinese corporate stakeholders informed us that they used its results to inform their investment decisions.

3.2. *Establishment of the Wildlife Refuge in the Area of Proposed Hydropower Development*

In parallel with the basin-wide assessment described above, the local government, WWF-Russia, the Rivers without Boundaries Coalition, and scientists undertook assessments and negotiations to establish a wildlife refuge. In 2015, the Verkhneamursky (Upper Amur) Wildlife Refuge, with an area of 239,639 ha, was established along the Shilka, Argun, and Amur rivers [62], covering three sites previously identified as suitable for the construction of large hydropower reservoirs. One of the proposed sites was the Transsibirskaya dam, as shown in Figure 7.

The new protected area covered the potential dam building sites on both the Shilka River and Upper Amur River, and specific regulations for this wildlife refuge establishment made the future planning and development of water infrastructure illegal. To open these sites for development, the provincial government would have to undertake specific painstaking bureaucratic procedures (e.g., assessments, public consultations) to modify the legal protection regime or boundaries of the protected area, which may also serve as an additional deterrent for potential investors.

Unlike strategic environmental assessments, the procedure of creating a wildlife refuge in Russia includes mandatory public consultations with local communities. Such discussions were held in the Mogochinsky District of Zabaikalsky Province in 2015 and resulted in modification of the intended protection regime to accommodate specific traditional uses by local communities, such as hunting and fishing. Given strong opposition against dam construction on the Shilka River among local people, as well as fears of forest devastation as a consequence of the Chinese-built Amazarsky pulp mill [63], the idea to establish a vast nature reserve along the major rivers gained wide popular support.

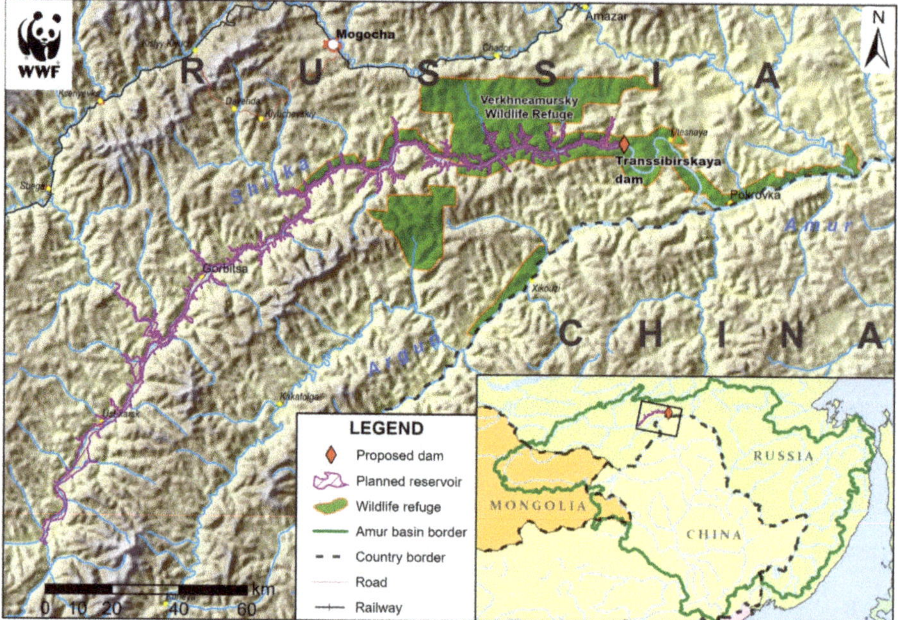

Figure 7. Map of the Verkhneamusky (Upper Amur) Wildlife Refuge and the potential Transsibirskaya dam site. © WWF-Russia.

3.3. Difficulties of Managing Environmental Flows in the Amur Basin

In the Amur basin, the main stakeholders pushing for the development of environmental releases from the already existing reservoirs are NGOs and scientists. Understanding the difficulties connected with environmental flow assessment and integration into hydropower management, NGOs have chosen a strategy to prevent initial damage instead of mitigating it after dam construction. Attempts to assess and develop environmental flows were made in 1993 [49], 2007–2010 [64], and 2013 [65]. However, as of 2019, the requirements for environmental flows from the reservoirs on the Zeya and Bureya rivers to sustain the freshwater and floodplain ecosystems below the dams had not been prescribed in the "Reservoir Operating Rules". Solving this problem is difficult due to the lack of knowledge of the biological resources of these freshwater ecosystems. In addition, for the Zeya River, there is no data on the initial conditions of the river ecosystem. This obstructs the assessment of the effectiveness of any implemented environmental flows. All these factors result in difficulties in assessing the damage caused by existing dams and developing adequate measures for its minimization and compensation.

4. Discussion

This section discusses how the work in the Amur basin compares to work in other basins and the possible reasons for notable variations in strategy, tactics, and outcomes. It explores aspects such as an assessment of alternatives to hydropower development, the timing of the strategic assessments and further implementation of their results, the protection of river ecosystems by establishing protected areas at undeveloped river stretches, environmental flows, and different strategies for their integration into reservoir management.

4.1. Amur and Baikal: Different Approaches to Alternatives

In different basins, the focus of key elements of strategic assessments could be quite different depending on local circumstances or changes in the global energy development context. In 2014–2018, members of the Rivers without Boundaries International Coalition (RwB) were involved in a basin-wide analysis of potential hydropower impacts in Lake Baikal and the Selenge River basin, shared by Mongolia and Russia. Mongolian authorities, supported by the World Bank, wanted to develop feasibility studies for two specific large dams, while the RwB members and local communities in Mongolia and Russia (as well as the World Heritage Committee) insisted that a strategic environmental assessment should first be carried out to determine the limitations for water infrastructure development in the basin and to explore available alternatives. In contrast to the analysis of alternatives in the Amur River basin, which focused on a comparison of impacts from different hydropower development scenarios, the emphasis of the analysis requested for the Lake Baikal basin was on a comparison of intended hydropower against alternative means to improve the energy supply system of Mongolia, including the deployment of solar, wind, and transboundary transmission grids, pumped storage, and other storage technologies. So far, Mongolia has agreed to conduct a regional environmental assessment for the Baikal–Selenge basin and has committed to undertake a thorough analysis of alternative technologies to the proposed hydropower dams [66]. This type of shift in focus from an analysis of dam siting to optimize hydropower development to considering alternative generating technologies at the energy system level is seen by the expert community as an important necessity (see, e.g., advice for donor governments [12], handbook on paradigm shifts in energy system planning [15], analysis of emerging trends in the Greater Mekong Subregion [67].

4.2. Basin-Wide Stragic Assessment: Experience in the Amur and Elsewhere

In the same decade that the WWF&En+ strategic assessment took place in Amur, many other basin-wide strategic environmental assessments (and other EIAs) have been carried out in different large transboundary basins [68]. Some of those assessments have been much more elaborate and detailed than the express analysis undertaken for Amur and were sanctioned and funded by intergovernmental organizations, governments of basin countries, donor agencies, and other relevant sources of authority. A detailed comparative study is needed in the future on the specific place and role of such SEA efforts in public education and decision-making processes [69]. Policy outcomes informed by environmental assessments vary dramatically. Shortcomings in the "Environmental Impact Report of Hydropower Development in the Upper Reaches of the Ayeyarwady River in Myanmar", performed by the Changjiang Institute of Survey, Planning, Design, and Research (CISPDR) in 2011 and subsequently highlighted by an independent review by International Rivers, contributed to the continued freezing of the controversial Myitsone dam [70]. Meanwhile, in another important case, a state-of-the-art presentation of overwhelming evidence on the negative consequences from damming the Mekong River did not stop those dams from being developed by Laos with investment from various international funders [71]. In a recent case in Myanmar, a key player, the Energy Ministry, in 2018 removed its endorsement from the final report of the most elaborate and balanced country-wide SEA, allegedly yielding to pressure from foreign firms with licenses for large hydropower development on major rivers [72]. Nevertheless, there are many examples from Vietnam, India, Pakistan, and Nepal where the recommendations of an SEA were not challenged and were partly used in decision-making. However, long-term perspectives and systemic findings of SEAs often contradict short-term interests of policy-makers and confront the shortcomings of projects already designed and invested in by developers.

The reason for the relative success of the Amur basin-wide assessment in helping to prevent further development of the most harmful hydropower projects is that it was undertaken at the earliest stages of planning, when potential project proponents were exploring project opportunities and had not yet invested heavily in a preferred option. Another important factor was the continuity of institutional knowledge and capacity building among Russian conservation NGOs and the expert community, who had, since 1990, preserved evidence on hydropower impacts that was collected during earlier assessments. Finally, little formal involvement from governmental bureaucratic mechanisms also likely contributed to the relative ease with which the assessment outcomes were internalized by various stakeholders. Even the pledge by the company to withdraw from the Transsibirskaya dam project on the Shilka River was a voluntary action without any legal obligations or dangerous consequences.

Through the activities of the Rivers without Boundaries International Coalition, the methodology and experiences from the Amur River assessment have been shared and thoroughly discussed with experts and CSOs from China, Australia, Netherlands, Myanmar, Thailand, Nepal, and Brazil. Various elements of the methodology could be applied to designing basin-wide assessments in other regions. However, even such a simple methodology requires indices to be modified for different river basins to represent the main impact factors. New factors may also need to be added, such as water withdrawal, which was very modest in the case of the Amur basin but may be significant in other regions.

4.3. Conservation of River Ecosystems by Protected Areas

The movement to set aside undeveloped river stretches [73] and preserve free-flowing rivers has a long history with a growing and diverse number of precedents and supporters [74,75]. This discussion will focus only on efforts to use traditional protected areas to protect river ecosystems in East Asia. In China, extensive efforts have been undertaken by civil society to prevent the damming of the Nu River (called Thanlwin or Salween downstream in Myanmar) and to preserve the last undammed stretches of the Yangtze–Jinsha River (e.g., Tiger Leaping Gorge). Both river stretches are immediately adjacent to the "Three Parallel Rivers of Yunnan PAs" World Heritage Site, which was explicitly designed to not include riverbanks to avoid conflict with planned hydropower [76]. Nevertheless, World Heritage status has been an important factor slowing and temporarily halting hydropower development in these stretches, but there are signs it may restart in the near future since no explicit "red lines" have been drawn to limit such development in perpetuity [77].

Another example is the Upper Yangtze Rare and Endemic Fish National Nature Reserve, the last undammed stretch in the Yangtze midflow, whose boundaries were redrawn twice in 2005 and in 2011 to make way for hydropower development [78]. The first revision excluded the most valuable habitats to give way for the giant Xiangjiaba and Xiluodu hydropower projects. However, the second revision of the reserve's borders, which allowed the construction of the Xiaonanhai dam and two more dams upstream of Chongqing City, was reversed in 2015 after wide public discussion initiated by Friends of Nature and other Chinese NGOs. The presence of the Endemic Fish National Nature Reserve has been central to the argument made by NGOs and is at the heart of the reasoning in the decision announced by the Ministry of Environmental Protection to further limit dam development [79]. The Endemic Fish National Nature Reserve, while not offering absolute protection, was still an important factor in the safeguarding from damming of the last undeveloped stretch of the Yangtze, but it has limited value in terms of preserving the biodiversity of migratory fish, since it is located on a fragmented river between large dams. For example, the cascade of dams blocking a spawning habitat has led to an ongoing decline in the number of adult Chinese sturgeon in the Yangtze River and the sea, from more than 32,000 before 1981 to 2500 in 2015, and extinction in the wild is possible within the next decade [80].

On the Amur River main channel, the development of the Taipinggou National Nature Reserve in China and Verkhneamursky (Upper Amur) Wildlife Refuge in Russia has provided a certain degree of protection to the two river stretches most susceptible to damming and has played an important role in an overall multifaceted effort to preserve the main channel of the Amur as a free-flowing river.

Taking into account the timing and dynamics of investors' decision-making has been an important factor in the success of this conservation effort. The creation of such protected areas has become possible because the intention of various stakeholders to develop large hydropower at a given site was not constant over long periods, but reemerged sporadically. Attempts to dam the Amur main stem were repeated in 1986–1993, 2000, 2007, and 2016, while during the periods in between, hydropower design was not a focus of interest for any investors or politicians. In addition, in Russia, the stakeholders who are likely to benefit the most from hydropower development are not local (e.g., hydropower companies, investors, national agencies), and the opportunity to build a dam is not seen as an essential economic opportunity by the majority of local communities or even local authorities. In contrast, in China, specific river stretches are formally assigned to large energy companies for development upon agreement with local municipalities, and this dramatically decreases the chances for successful gazetting of protected areas at such sites with the consent of local authorities.

Since freshwater biodiversity faces a severe crisis driven by impacts from large infrastructure, this approach is considered highly valuable, and work continues with CSOs globally to make business, governments, and intergovernmental bodies recognize it as an important and urgent measure to protect free-flowing rivers [29].

4.4. Compensatory Measures to Reduce Dam Impacts on Freshwater Ecosystems

The best tactic to protect fragile freshwater ecosystems is to avoid harmful hydropower construction by conducting a strategic environmental assessment, maintaining free-flowing rivers, and proposing careful dam siting, design, and operation. For existing dams or dams under construction, measures to minimize and compensate for their negative impacts to maintain and restore ecosystems should be implemented. Downstream effects of dams can be minimized through the provision of environmental flows and fish passages. Ecosystem restoration can be achieved through improved species, habitat, or catchment management interventions. Compensation measures implemented at different hydropower projects tend to be more effective when planned at an early stage with a basin-wide view of the situation.

On the Russian tributary of the Amur River, a program for ecosystem conservation was undertaken alongside the construction of the Lower Bureya dam (320 MW). The "Bureisky Compromise" environmental program was implemented by the hydropower company RusHydro, the local government, the UNDP implementing a project of the Global Environmental Fund (UNDP/GEF), and ecologists. The project activities were aimed at minimizing the negative impact of the Lower Bureya dam's construction on the biodiversity of the terrestrial ecosystem. In Russia, this was the first time such an environmental compensation program had been used when constructing a large dam [81]. However, the program did not include measures aimed at conserving freshwater ecosystems or species, and no study was undertaken to design environmental flow releases. As a result, the feasibility for freshwater ecosystem conservation of the Lower Bureya and Middle Amur was not assessed.

Examples of reportedly successful dam reoperation with environmental flow releases are known from several rivers of the world, including from the Three Gorges dam on the Yangtze River, China [82]. These releases, which have been provided since 2011, are successfully contributing to carp spawning [83]. However, potential improvements that could be achieved on the Yangtze via introducing environmental flows may offset only a small part of the negative impacts on freshwater biodiversity caused by extensive hydropower development. Another example of a high-resolution environmental flow assessment is the 102-MW Gulpur Hydropower Project on the Poonch River (Pakistan). The assessment resulted in decisions such as (1) reducing the dewatered section by relocating the weir closer to the powerhouse, (2) releasing environmental flows, (3) implementing a management plan for the Poonch River National Park, and (4) establishing a fish hatchery used to stock the reach downstream of the Gulpur tailrace with native fish. Additionally, the construction of two more dams on the Poonch River was cancelled due to incompatibility with the preservation of the river ecosystem in the National Park [9].

5. Conclusions

Hydropower and other water infrastructure development have caused a dramatic worldwide decline in the number of connected, free-flowing rivers due to haphazard planning and disregard for environmental and social values [5]. Clear limits should be placed on the allowable alteration of a large river system by water infrastructure development so that a basin can retain its vital natural processes, species diversity and abundance, vital ecosystem services, and associated cultural values.

Long-term conservation of freshwater ecosystems can only be achieved on a system scale. In most cases, a strategic assessment of hydropower has to be performed at the basin level to be successful, rather than being limited to the project implementation site, although this has certain trade-offs in the ability to achieve all-inclusive public participation. Stakeholder analysis and proper engagement are as important for environmental assessments as scientific research. Public participation has been the key process in hydropower development, even when no formal public consultation procedure was implemented by dam project proponents.

To mitigate potential conflicts emerging at the Transsibirskaya dam, WWF-Russia and En+ Group launched a basin-wide research project at the transboundary Amur basin. The ultimate goal of the strategic assessment was to evaluate different options and identify hydropower dam location options with the least environmental impacts for the whole basin, as well as maximize the social and economic benefits to the region. To achieve this, a new methodology for the rapid environmental assessment of hydropower development in large river basins was developed. This methodology can be applied before any investments or construction takes place and is able to assess and rank options using three main factors that determine the environmental impacts of hydropower.

As a result of the project and despite the expected economic benefits from the Transsibirskaya dam, the En+ Group company stated that the joint assessment showed that a dam on the Shilka River would be an unbalanced/unsustainable option due to its environmental and social impacts. The company and its Chinese counterparts have not pursued the development of this dam any further. At the same time, development of the Lower Bureya hydropower dam proceeded smoothly without major conflicts with environmental CSOs because it was characterized by low expected impacts on the freshwater ecosystem of the Amur. Thus, the design of a common roadmap with the expectations and aspirations of both energy companies and civil society groups decreased conflict and allowed stakeholders to concentrate on identifying mutually acceptable development options.

Based on this experience, the timing and sequence of the assessment tools used is the key factor in mitigating impacts from hydropower development. Starting earlier and fully incorporating the perspectives of affected stakeholders improves the chances that optimal decisions are made and that the least negative impacts are suffered by companies, communities, and natural ecosystems. The effectiveness of initial avoidance and mitigation of hydropower impacts at early planning stages is far higher compared to applying measures after dam construction.

Applying the newly developed methodology, which includes all possible development scenarios combining existing and potential large dams in the Amur basin, it was concluded that maintenance of the free-flowing Shilka River is essential for sustaining the ecological health of the Amur basin as a whole, and the Transsibirskaya dam scored among five of the most environmentally damaging development scenarios.

Other results of the research illustrated the following:

1. Even if a hydropower scheme is carefully planned, a variety of relatively low-impact scenarios are available only for harnessing the first 10–25% of basin-wide hydropower potential. In any scenario attempting to realize a greater proportion of the technically feasible potential of the Amur basin, the expected environmental impacts grow dramatically, and the difference between the "best" and the "worst" scenarios becomes insignificant. No sustainable development options were found for utilizing the majority of basin-wide hydropower potential;
2. The main factor limiting opportunities for future sustainable development of hydropower in the Amur basin is the negative cumulative impact resulting from existing dams. If planned in an environmentally sound way, the current generating capacity could have been developed with an environmental impact two times smaller;
3. A hydropower cascade on the main stem of the Amur River would be associated with the highest basin-wide environmental impact (which is consistent with findings from many other large basins), while additional dams on tributaries already altered by hydropower are associated with the smallest additional basin-wide environmental impact.

5.1. Long-Term Protection of Rivers and Ecosystems

The creation of protected areas in places suitable for damming can constrain hydropower development at ecologically sensitive sites and lead to long-term protection of wild rivers and surrounding terrestrial ecosystems. The chances of success depend on the timing of the nature reserve planning vis-à-vis the hydropower project cycle and the attitudes toward hydropower held by the local population and authorities responsible for protected area development. The threat of hydropower, which certainly blocks access to many traditionally used resources, often causes local communities to proactively support nature reserve creation, despite the restrictions it imposes on land use. This approach has led to the successful establishment of three nature reserves in places suitable for hydropower dams in the Amur River basin and thus should be considered replicable at least in this large basin. Efforts to mitigate dam impacts in the Amur basin were more successful than in the Yangtze basin for a variety of reasons, such as modest prospects for hydropower development in the Amur basin, fewer obstacles in the legal system in Russia, and the difference in the hydropower development process in Russia and China. Nevertheless, the general utility of this approach has been demonstrated in the Yangtze River basin as well, albeit with less obvious long-term outcomes for basin-wide ecosystem conservation.

5.2. Minimization of and Compensation for Negative Impacts

For existing dams or dams under construction, measures to minimize and compensate for negative impacts and maintain and restore the affected ecosystems should be implemented. Downstream impacts can be minimized through the provision of environmental flows. Currently, requirements for environmental flow releases for the reservoirs on the Zeya and Bureya rivers have not been developed. In Russia, environmental flow releases are not implemented in practice for most reservoirs. Since the legislative framework does not force water management stakeholders to implement measures for the conservation and restoration of freshwater ecosystems, it is difficult to assess the damage caused by existing dams and to develop adequate means for its minimization and compensation. Another reason for the lack of integration of environmental flow releases is the lack of influence and interest on the part of societal groups who could potentially benefit from them. Another problem is the lack of proper monitoring and the shortage of data on the initial state of the ecosystem, without which it is difficult to assess the effectiveness of the environmental flow regime if it is implemented.

5.3. Building Additional Tools

The conservation methodologies developed and employed so far should be complemented in the Amur basin by detailed mapping of the distribution and connectivity between biodiversity hotspots [84]. As the first step in this work, the "Fishes of the Amur" Atlas was published [85]. The underlying database on the distribution and abundance of fish species will be used in the future to supplement basin-wide rapid assessments with a component taking into account the distribution and migration of fish species.

All conservation tools should be used in an interrelated manner in one robust planning process. To produce a conservation and development master plan for the Amur aimed at sustaining a healthy free-flowing river ecosystem, a comprehensive strategic environmental assessment of the current river basin management system and various plans for future development (including hydropower) is necessary. Of course, this arduous task is complicated by the division of our basin between China, Russia, and Mongolia, but must be undertaken to build a new "ecological civilization" together and achieve harmony between nature and humans.

Supplementary Materials: The following are available online at http://www.mdpi.com/2073-4441/11/8/1570/s1, Figure S1: Delineated 214 units/stretches with attributed characteristic of respective elementary sub-basins, Table S1: Calculation of the main factors of dam impact by river sections, Table S2: Primary data on hydropower dams and rivers for the production of calculations on the analyzed sections.

Author Contributions: Conceptualization, E.A.S.; Methodology, E.A.S. and E.G.E.; Software, E.G.E.; Validation, E.A.S., E.G.E., and O.I.N.; Formal Analysis, E.A.S.; Investigation, E.A.S. and E.G.E.; Resources, E.A.S. and E.G.E.; Data Curation, E.G.E.; Writing-Original Draft Preparation, E.A.S.; Writing-Review & Editing, O.I.N.; Visualization, E.G.E. and O.I.N.; Supervision, O.I.N.; Project Administration, O.I.N.; Funding Acquisition, E.A.S.

Funding: This research was funded by World Wide Fund for Nature (WWF-Russia) grant number WWF001281/RU009627-19ru/GLM, the Whitley Fund for Nature grant number 3, and the Whitley Fund for Nature supporting "Keep the Rivers Free and Wild" Project coordinated by the Rivers without Boundaries Coalition.

Conflicts of Interest: The authors declare no conflict of interest.

References

1. Balian, E.V.; Segers, H.; Martens, K.; Lévéque, C. The Freshwater Animal Diversity Assessment: An overview of the results. *Hydrobiologia* **2008**, *595*, 627–637. [CrossRef]
2. Poff, N.L.; Allan, J.D.; Bain, M.B.; Karr, J.R.; Prestegaard, K.L.; Richter, B.D.; Sparks, R.E.; Stromberg, J.C. The natural flow regime: A paradigm for river conservation and restoration. *BioScience* **1997**, *47*, 769–784. [CrossRef]
3. Karr, J.R. Assessment of Biotic Integrity Using Fish Communities. *Fisheries* **1981**, *6*, 21–27. [CrossRef]
4. LPR. *Living Planet Report 2018: Aiming Higher*; Grooten, M., Almond, R.E.A., Eds.; WWF: Gland, Switzerland, 2018; p. 35. Available online: https://www.worldwildlife.org/pages/living-planet-report-2018 (accessed on 10 July 2019).
5. Grill, G.; Lehner, B.; Thieme, M.; Geenen, B.; Tickner, D.; Antonelli, F.; Babu, S.; Borrelli, P.; Cheng, L.; Crochetiere, H.; et al. Mapping the world's free-flowing rivers. *Nature* **2019**, *569*, 215–221. [CrossRef] [PubMed]
6. ICOLD (International Commission on Large Dams). World Register of Dams. 2017. Available online: www.icold-cigb.net/GB/world_register/general_synthesis.asp (accessed on 10 July 2019).
7. Arthington, A.H.; Bhaduri, A.; Bunn, S.E.; Jackson, S.E.; Tharme, R.E.; Tickner, D.; Young, B.; Acreman, M.; Baker, N.; Capon, S.; et al. The Brisbane Declaration and Global Action Agenda on Environmental Flows. *Front. Environ. Sci.* **2018**, *6*, 45. [CrossRef]
8. Moran, E.F.; Lopez, M.C.; Moore, N.; Müller, N.; Hyndman, D.W. Sustainable hydropower in the 21st century. *Proc. Natl. Acad. Sci. USA* **2018**, *115*, 11891–11898. [CrossRef] [PubMed]
9. World Bank Group. *Environmental Flows for Hydropower Projects: Guidance for the Private Sector in Emerging Markets*; International Finance Corporation: Washington, DC, USA, 2018. Available online: https://openknowledge.worldbank.org/handle/10986/29541 (accessed on 10 July 2019).
10. IHA (International Hydropower Association). *Hydropower Sustainability Assessment Protocol*; IHA: London, UK, 2010. Available online: http://www.hydrosustainability.org/Hydropower-Sustainablility-Assessment-Protocol/Documents.aspx (accessed on 10 July 2019).
11. IHA (International Hydropower Association). Do the Benefits of Sustainable Hydropower Outweigh the Costs? 13 March 2014. Available online: https://www.hydropower.org/blog/do-the-benefits-of-sustainable-hydropower-outweigh-the-costs (accessed on 10 July 2019).
12. Opperman, J.; Hartmann, J.; Lambrides, M.; Carvallo, J.P.; Chapin, E.; Baruch-Mordo, S.; Eyler, B.; Goichot, M.; Harou, J.; Hepp, J.; et al. *Connected and Flowing: A Renewable Future for Rivers, Climate and People*; WWF and The Nature Conservancy: Washington, DC, USA, 2019; p. 55. Available online: https://www.wwf.de/fileadmin/fm-wwf/Publikationen-PDF/Connected-and-Flowing.pdf (accessed on 10 July 2019).
13. Liu, Z. *Global Energy Interconnection*; Academic Press: Oxford, UK, 2015; p. 396, ISBN 978-0-12-804405-6.
14. NCEA (Netherlands Commission for Environmental Assessment). *ESIA and SEA for Sustainable Hydropower Development*; Views and Experiences Series; Netherlands Commission for Environmental Assessment: Utrecht, The Netherlands, 2018. Available online: https://www.eia.nl/documenten/00000339.pdf (accessed on 10 July 2019).
15. NCEA (Netherlands Commission for Environmental Assessment); TNC (The Nature Conservancy). A Strategic Approach to Hydropower Development. Unpublished manuscript. 2019. Available online: https://www.eia.nl/docs/mer/diversen/draft_advocacy_paper_5.7.19.pdf (accessed on 10 July 2019).
16. NCEA (Netherlands Commission for Environmental Assessment). *Better Decision-Making about Large Dams with a View to Sustainable Development*, 2nd ed.; Policy Document (7199); Dutch Sustainability Unit: Utrecht, Dutch, 2017; p. 21. Available online: http://www.rainfoundation.org/news/better-decision-making-large-dams-view-sustainable-development/ (accessed on 10 July 2019).

17. WCD (World Commission on Dams). *Dams and Framework for Decision-Making*; Earthscan Publications Ltd.: London, UK, 2000; ISBN 1-85383-798-9.
18. Tullos, D.; Tilt, B.; Liermann, C.R. Introduction to the special issue: Understanding and linking the biophysical, socioeconomic and geopolitical effects of dams. *J. Environ. Manag.* **2009**, *90*, S203–S207. [CrossRef] [PubMed]
19. Gao, J.; Kørnøv, L.; Christensen, P. The politics of strategic environmental assessment indicators: Weak recognition found in Chinese guidelines. *Impact Assess. Proj. Apprais.* **2013**, *31*, 232–237. [CrossRef]
20. Lee, Y.B. Water Power: The "Hydropower Discourse" of China in an Age of Environmental Sustainability. *ASIANetwork Exch. A J. Asian Stud. Lib. Arts* **2014**, *21*, 42–51. [CrossRef]
21. Fearnside, P.M. Brazil's São Luiz do Tapajós dam: The art of cosmetic Environmental Impact Assessments. *Water Altern.* **2015**, *8*, 373–396.
22. Maksimov, A.A. *Hydropower Projects on the Traditional Lands of the Indigenous Peoples of the North: International Standards and Practice*; Library of Indigenous Peoples of the North: Moscow, Russia, 2010; Volume 15.
23. Yeophantong, P. China's Lancang Dam Cascade and Transnational Activism in the Mekong Region. *Asian Surv.* **2014**, *54*, 700–724. [CrossRef]
24. Foran, T.; Kiik, L.; Hatt, S.; Fullbrook, D.; Dawkins, A.; Walker, S.; Chen, Y. Large hydropower and legitimacy: A policy regime analysis, applied to Myanmar. *Energy Policy* **2017**, *110*, 619–630. [CrossRef]
25. IFC (International Finance Corporation). Strategic environmental assessment (SEA) of the hydropower sector in Myanmar. In *Baseline Assessment Report*; Peace and Conflict; International Finance Corporation: Washington, DC, USA, 2017.
26. Mahato, B.K.; Ogunlana, S.O. Conflict dynamics in a dam construction project: A case study. *Built Environ. Proj. Asset Manag.* **2011**, *1*, 176–194. [CrossRef]
27. Abell, R.; Thieme, M.L.; Revenga, C.; Bryer, M.; Kottelat, M.; Bogutskaya, N.; Coad, B.; Mandrak, N.; Balderas, S.C.; Bussing, W.; et al. Freshwater Ecoregions of the World: A New Map of Biogeographic Units for Freshwater Biodiversity Conservation. *BioScience* **2008**, *58*, 403–414. [CrossRef]
28. Olson, D.M.; Dinerstein, E.; Wikramanayake, E.D.; Burgess, N.D.; Powell, G.V.N.; Underwood, E.C.; D'amico, J.A.; Itoua, I.; Strand, H.E.; Morrison, J.C.; et al. Terrestrial Ecoregions of the World: A New Map of Life on Earth. *BioScience* **2001**, *51*, 933–938. [CrossRef]
29. Dye, B.J.; Simonov, E.; Lafitte, G.; Bennett, K.; Scarry, D.; Lepcha, T.; Lepcha, G.; Wagh, S.; Kemman, A.; Eyler, B.; et al. *Heritage Dammed: Water Infrastructure Impacts on World Heritage Sites and Free Flowing Rivers*; Rivers without Boundaries, World Heritage Watch: Moscow, Russia, 2019; p. 132, ISBN 978-5-4465-2345-0.
30. Gusev, M.N.; Pomiguev, Y.V. Channel activity of trunk rivers of the Amurskaya region in conditions of current economic management. *Geogr. Nat. Resour.* **2008**, *29*, 137–141. [CrossRef]
31. Podolsky, S.A.; Simonov, E.A.; Darman Yu, A. *Where Does the Amur Flow?* WWF-Russia: Moscow, Russia, 2006; p. 72.
32. Semenchenko, N.N. Hydrological regime of the Amur River and the number of commercial freshwater fish. In *The Current State of Aquatic Bioresources: Materials of Scientific. Conf., Dedicated to the 70th Anniversary of S.M. Konovalov*; TINRO-Center: Vladivostok, Russia, 2008; pp. 246–250.
33. Kotsyuk, D.V. Formation of ichthyofauna of the Zeya Reservoir: A retrospective analysis and current status. In *Dissertation*; TINRO-Center: Khabarovsk, Russia, 2009.
34. Zausaev, V.K.; Khaliullina, Z.A.; Goryainov, V.A.; Sirotsky, S.E.; Gorbunov, N.M. *Monitoring the Social and Economic Impact of the Bureyskaya HPP Zone*; Zausaev, V.K., Ed.; Institute of Water and Ecological Problem of the Far Eastern Branch of the Russian Academy of Sciences: Khabarovsk, Russia, 2007; p. 126, ISBN 5-7442-1440-2.
35. Department of Science and Technology Policy and Development of RAO UES of Russia. *Guidelines for Assessing the Impact of Hydraulic Structures on the Environment*; Regulatory technical document. RD 153-34.2-02.409-2003. Approved by the Department of science and technology policy and development of RAO UES of Russia on January 24; Department of Science and Technology Policy and Development of RAO UES of Russia: Moscow, Russia, 2003.
36. Yi, Y. Recent Chinese policies on river hydropower plans (RHPS) and environmental impact assessments. In Proceedings of the 7th International Science Practical "The Rivers of Siberia and the Far East" Conference Materials, Khabarovsk, Russia, 30–31 May 2012; WWF-Russia: Moscow, Russia, 2012; pp. 202–205. Available online: https://wwf.ru/upload/iblock/f70/sbornik_rekisibiri.pdf (accessed on 10 July 2019).
37. White Book on Dams and Development in Russia. Available online: https://solex-un.ru/dams (accessed on 10 July 2019).

38. Opperman, J.J.; Harrison, D. *Pursuing Sustainability and Finding Profits: Integrated Planning at the System Level*; Hydrovision: Sacramento, CA, USA, 2008; p. 20.
39. Nilsson, C. Fragmentation and Flow Regulation of the World's Large River Systems. *Science* **2005**, *308*, 405–408. [CrossRef]
40. Vörösmarty, C.J.; McIntyre, P.B.; Gessner, M.O.; Dudgeon, D.; Prusevich, A.; Green, P.; Glidden, S.; Bunn, S.E.; Sullivan, C.A.; Liermann, C.R.; et al. Global threats to human water security and river biodiversity. *Nature* **2010**, *467*, 555–561. [CrossRef]
41. Richter, B.D.; Thomas, G.A. Restoring Environmental Flows by Modifying Dam Operations. *Ecol. Soc.* **2007**, *12*. [CrossRef]
42. Sapaev, V.M. Regulation of the Amur River. Is it possible to optimize environmental conditions? *Sci. Nat. FE* **2006**, *2*, 86–95.
43. Kotsyuk, D.V. Ichthyofauna structure and dynamics of the stock of basic food fish of the Zeya reservoir. In *Readings in Memory of S.M. Konovalov*; TINRO-Center: Vladivostok, Russia, 2008; pp. 133–137.
44. Simonov, E.A.; Egidarev, E.G.; Menshikov, D.A.; Khalyapin, L.E.; Korolyov, G.S. *Comprehensive Environmental and Socio-Economic Assessment of Hydropower Development in the Amur River Basin*; WWF-Russia, En+ Group: Moscow, Russia, 2015; p. 279. Available online: https://wwf.ru/resources/publications/booklets/comprehensive-environmental-and-socio-economic-assessment-of-hydropower-development-in-the-amur-rive/ (accessed on 10 July 2019).
45. Jarvis, A.; Reuter, H.I.; Nelson, A.; Guevara, E. Hole-Filled SRTM for the Globe Version 4, Available from the CGIAR-CSI SRTM 90m Database. 2008. Available online: http://srtm.csi.cgiar.org (accessed on 10 July 2019).
46. USGS (U.S. Geological Survey). Shuttle Radar Topography Mission Water Body Dataset. 2006. Available online: https://www.usgs.gov/centers/eros/science/usgs-eros-archive-digital-elevation-shuttle-radar-topography-mission-water-body?qt-science_center_objects=0#qt-science_center_objects (accessed on 22 May 2019).
47. Voznesensky, A. *Hydropower Energy Resources of the USSR*; USSR Academy of Sciences: Moscow, Russia, 1967; p. 312.
48. CAI, Chinese Academy of Engineering. *On Some Strategic Questions in Water and Land Resource Allocation, Environment and Sustainable Development in North East China*; Summary, Report; Fang, S.G., Ed.; Chinese Science Publishing: Beijing, China, 2007.
49. Ladygin, V.F.; Zhou, D. *Joint Russian-Chinese Comprehensive Scheme for Water Development in Transboundary Waters of the Argun and Amur Rivers*; Detailed Project Documentation (Vols. I–XIV); Sovintervod, Songliaowei: Moscow, Russia; Changchun, China, 1993.
50. Ladygin, V.F.; Zhou, D. *Joint Russian-Chinese Comprehensive Scheme for Water Development in Transboundary Waters of the Argun and Amur Rivers Synopsis*; Sovintervod, Songliaowei: Moscow, Russia; Changchun, China, 1999.
51. Ministry of Energy. *13th 5-Year Plan for Hydropower Development in China*; Ministry of Energy: Beijing, China, 2016. Available online: http://www.nea.gov.cn/2016-11/29/c_135867663.htm (accessed on 10 July 2019).
52. Glazyrina, I.P.; Fattakhov, R.V.; Delyuga, A.V.; Stroev, P.V.; Grigorov, A.A. *The Baikal Region: "Environmental Cost" of Economic Growth*; Regional Research of Russia: Novosibirsk, Russia, 2018; pp. 231–249.
53. Russian-Chinese Commission. *Sino-Russian Strategy for Transboundary System of Protected Areas in Amur River Basin in 2010–2020*; Annex 4 to the Protocol of the sixth session of the Sub-Commission on cooperation in the field of environmental protection of the Russian-Chinese Commission on preparation of regular meetings of Heads of Government; Russian-Chinese Commission: Harbin, China, 2011; p. 24, (In Chinese and Russian).
54. Gafarov, Y.M. *Nature Protected Areas of the Amur Region (Handbook)*; WWF-Russia: Vladivostok, Russia, 2013; p. 88.
55. Tharme, R.E. A global perspective on environmental flow assessment: Emerging trends in the development and application of environmental flow methodologies for rivers. *River Res. Appl.* **2003**, *19*, 397–441. [CrossRef]
56. Ministry of Natural Resources and Environment of the Russian Federation. *Order of the Ministry of Natural Resources and Environment of the Russian Federation of January 26 No. 17 "On approval of Guidelines for the Development of Rules for the Use of Reservoirs"*; Ministry of Natural Resources and Environment of the Russian Federation: Moscow, Russia, 2011.
57. Dubinina, V.G. *Methodical Foundations of the Environmental Regulation of the Water Irrevocable Withdrawals and the Establishment of Environmental Flow (Release)*; Economics and Informatics: Moscow, Russia, 2001; p. 120.

58. Trans-Sibirskaya HPP: Russia; EJAtlas: Barcelona, Spain. 2012. Available online: https://ejatlas.org/conflict/trans-sibirskaya-hpp-russia (accessed on 10 July 2019).
59. NGOs Meet with the China Three Gorges Corporation on Russian Projects. Available online: http://www.transrivers.org/2014/1105/ (accessed on 10 July 2019).
60. Simonov, E.A.; Nikitina, O.I.; Egidarev, E.G. River flood adaptation in the Amur Basin and nature conservation. *Use Prot. Nat. Resour. Russ. Bull.* **2015**, *3*, 15–24.
61. Simonov, E.A.; Nikitina, O.I.; Osipov, P.O.; Egidarev, E.G.; Shalikovsky, A.V. *Amur Floods and Us: Lesson (Un)Learned*; Shalikovsky, A.V., Ed.; WWF-Russia: Moscow, Russia, 2016; p. 216.
62. Korsun, O.V. *Verkhneamursky Reserve—The Pearl of the Mogochinsk District*, Educational ed.; House of Peace LLC: Novosibirsk, Russia, 2018.
63. Banktrack. Amazar Pulp and Saw Mill, Russia. Available online: https://www.banktrack.org/project/amazarsky_pulp_and_paper_mill (accessed on 10 July 2019).
64. Golovko, V.I. *Justification of the Need to Provide Fishery Releases in the Regulation of the Operation of Hydroelectric Power Plants of the Amur Region*; Appendix to the letter to the Ministry of Agriculture and Agriculture; Center for Fisheries Research: Blagoveshchensk, Russia, 2008.
65. Nikitina, O.I. Conclusion on the draft standards for allowable limits of impact on water bodies in the Amur River basin: The Argun. In *Complex Management and Protection of Water Resources. Opportunities to Optimize SIUPWR and SAL Mechanisms*; WWF-Russia: Moscow, Russia, 2013; pp. 23–40. Available online: https://amurinfocenter.org/upload/iblock/89e/skiovo-i-ndv.pdf (accessed on 10 July 2019).
66. Shuren Hydropower Plant Project, Mongolia. Available online: http://minis.mn/en/p/shuren-hydropower-plant-project?search=regional+environmental+assessment# (accessed on 10 July 2019).
67. Weatherby, C.; Eyler, B. *Mekong Power Shift: Emerging Trends in the GMS Power Sector*; Letters from the Mekong; The Stimson Center: Washington, DC, USA, 2017; p. 60. Available online: https://www.stimson.org/content/letters-mekong-mekong-power-shift-emerging-trends-gms-power-sector (accessed on 10 July 2019).
68. NCEA (Netherlands Commission for Environmental Assessment). Sustainable Hydropower Development: The Role of EIA and SEA: Utrecht, The Netherlands. 2015. Available online: https://www.commissiemer.nl/docs/mer/diversen/case_hydropwer_jan_2015.pdf (accessed on 10 July 2019).
69. Kolhoff, A. SEA—Hydropower sector. In Proceedings of the Hydroelectric Dams: Solution or Obstacle for Delivering the Paris Climate Agreement and Sustainable Development Goals, Paris, France, 13 May 2019.
70. Independent Expert Review Table of Contents of the Myitsone Dam EIA. In *International Rivers, ECODEV*; International Rivers: Oakland, CA, USA, 2013; p. 16. Available online: https://www.internationalrivers.org/sites/default/files/attached-files/independent_expert_review_of_the_myitsone_dam_eia_0.pdf (accessed on 10 July 2019).
71. ICEM (International Centre for Environmental Management). *Strategic Environmental Assessment of Hydropower on the Mekong Mainstream*; Summary of the Final Report; ICEM, International Centre for Environmental Management: Hanoi, Viet Nam, 2010. Available online: http://www.mrcmekong.org/assets/Publications/Consultations/SEA-Hydropower/SEA-FR-summary-13oct.pdf (accessed on 10 July 2019).
72. IFC (International Finance Corporation). *Strategic Environmental Assessment of the Myanmar Hydropower Sector*; Final Report; IFC, International Finance Corporation: Washington, DC, USA, 2018. Available online: https://www.ifc.org/wps/wcm/connect/2f7c35f4-e509-48b2-9fd8-b7cbc0501171/SEA_Final_Report_English_web.pdf?MOD=AJPERES (accessed on 10 July 2019).
73. United States Fish and Wildlife Service. National Wild and Scenic Rivers System: About the WSR Act. 2016. Available online: www.rivers.gov/wsr-act.php (accessed on 10 July 2019).
74. WWF (World Wide Fund for Nature). *Free-Flowing Rivers—Economic Luxury or Ecological Necessity?* WWF International: Gland, Switzerland, 2006. Available online: https://wwf.panda.org/?63020/Free-flowing-rivers-Economic-luxury-or-ecological-necessity (accessed on 10 July 2019).
75. Moir, K.; Thieme, M.L.; Opperman, J. *Securing a Future That Flows: Case Studies of Protection Mechanisms for Rivers*; World Wildlife Fund for Nature and The Nature Conservancy: Washington, DC, USA, 2016.
76. Lafitte, G. Great Leaping Tiger Dammed. In *Heritage Dammed: Water Infrastructure Impacts on World Heritage Sites and Free Flowing Rivers, Civil Society Report to the UNESCO World Heritage Committee and Parties of the World Heritage Convention*; Rivers without Boundaries and World Heritage Watch: Moscow, Russia, 2019.

77. UNESCO. World Heritage Committee Decision: 41 COM 7B.27. Three Parallel Rivers of Yunnan Protected Areas (China) (N 1083bis). 2017. Available online: https://whc.unesco.org/en/decisions/7029 (accessed on 10 July 2019).
78. Zhang, B. Xiaonanhai, elegy of the natural reserve in the new round of river development. In Proceedings of the 7th International Science Practical "The Rivers of Siberia and the Far East" Conference Materials, WWF-Russia, Khabarovsk, Russia, 30–31 May 2012; pp. 211–217. Available online: https://wwf.ru/upload/iblock/f70/sbornik_rekisibiri.pdf (accessed on 10 July 2019).
79. Ministry of Environmental Protection. *Ministry of Environmental Protection's Wudongde Approval Letter (includes Rejection Decision of Xiaonanhai Dam)*; Ministry of Environmental: Beijing, China, 2015. Available online: https://www.internationalrivers.org/sites/default/files/attached-files/wu_dong_de_dian_zhan_huan_ping_pi_fu_.pdf (accessed on 10 July 2019). (In Chinese)
80. Huang, Z.; Wang, L. Yangtze Dams Increasingly Threaten the Survival of the Chinese Sturgeon. *Curr. Biol.* **2018**, *28*, 3640–3647. [CrossRef] [PubMed]
81. RusHydro. At the Nizhne-Bureiskaya HPP, the "Bureysky Compromise" Program is Being Completed. 2017. Available online: http://www.rushydro.ru/press/news/103941.html (accessed on 10 July 2019).
82. Harwood, A.; Johnson, S.; Richter, B.; Locke, A.; Yu, X.; Tickner, D. *Listen to the River: Lessons from a Global Review of Environmental Flow Success Stories*; WWF-UK: Working, UK, 2017. Available online: https://www.wwf.org.uk/sites/default/files/2017-09/59054%20Listen%20to%20the%20River%20Report%20download%20AMENDED.pdf (accessed on 10 July 2019).
83. Cheng, L.; Opperman, J.J.; Tickner, D.; Speed, R.; Guo, Q.; Chen, D. Managing the Three Gorges Dam to Implement Environmental Flows in the Yangtze River. *Front. Environ. Sci.* **2018**, *6*, 64. [CrossRef]
84. Osipov, P.E.; Egidarev, E.G.; Rydannykh, A.O.; Mikheev, I.E.; Simonov, E.A. GIS applications for delineation of aquatic habitats in need of protection in the Amur River Basin. *Amur. Zool. J.* **2015**, *7*, 277–284. (In Russian)
85. Antonov, A.L.; Barabanshikov, E.I.; Zolotukhin, S.F.; Mikheev, I.E.; Shapovalov, M.E. *Fishes of the Amur*; World Wide Fund for Nature (WWF): Vladivostok, Russia, 2019; p. 318.

Article

Copula-Based Research on the Multi-Objective Competition Mechanism in Cascade Reservoirs Optimal Operation

Menglong Zhao [1], Shengzhi Huang [1,*], Qiang Huang [1], Hao Wang [2], Guoyong Leng [3], Siyuan Liu [1] and Lu Wang [1]

1 State Key Laboratory Base of Eco-Hydraulic Engineering in Arid Area, Xi'an University of Technology, Xi'an 710048, China; mlzhao1314@163.com (M.Z.); syhuangqiang@163.com (Q.H.); syl1336@outlook.com (S.L.); wlbless@163.com (L.W.)

2 State Key Laboratory of Simulation and Regulation of Water Cycle in River Basin, China Institute of Water Resources and Hydropower Research, Beijing 100038, China; ziyuanxxb@126.com

3 Environmental Change Institute, University of Oxford, Oxford OX1 3QY, UK; guoyong.leng@gmail.com

* Correspondence: huangshengzhi@xaut.edu.cn

Received: 10 April 2019; Accepted: 9 May 2019; Published: 12 May 2019

Abstract: Water resources systems are often characterized by multiple objectives. Typically, there is no single optimal solution which can simultaneously satisfy all the objectives but rather a set of technologically efficient non-inferior or Pareto optimal solutions exists. Another point regarding multi-objective optimization is that interdependence and contradictions are common among one or more objectives. Therefore, understanding the competition mechanism of the multiple objectives plays a significant role in achieving an optimal solution. This study examines cascade reservoirs in the Heihe River Basin of China, with a focus on exploring the multi-objective competition mechanism among irrigation water shortage, ecological water shortage and the power generation of cascade hydropower stations. Our results can be summarized as follows: (1) the three-dimensional and two-dimensional spatial distributions of a Pareto set reveal that these three objectives, that is, irrigation water shortage, ecological water shortage and power generation of cascade hydropower stations cannot reach the theoretical optimal solution at the same time, implying the existence of mutual restrictions; (2) to avoid subjectivity in choosing limited representative solutions from the Pareto set, the long series of non-inferior solutions are adopted to study the competition mechanism. The premise of sufficient optimization suggests a macro-rule of 'one falls and another rises,' that is, when one objective value is inferior, the other two objectives show stronger and superior correlation; (3) the joint copula function of two variables is firstly employed to explore the multi-objective competition mechanism in this study. It is found that the competition between power generation and the other objectives is minimal. Furthermore, the recommended annual average water shortage are 1492×10^4 m^3 for irrigation and 4951×10^4 m^3 for ecological, respectively. This study is expected to provide a foundation for selective preference of a Pareto set and insights for other multi-objective research.

Keywords: multi-objective competition mechanism; cascade reservoirs operation; copula function; Pareto set

1. Introduction

Real-world systems always refer to multiple objective optimization in their operations. Multi-objective optimization problems (MOPs) usually require the simultaneous optimization of some incommensurable and competitive objectives. A reservoir system, for example, serves various purposes and involves multi-objective decision-making in the implementation process. With continuous development and utilization of water resources and hydropower resources and the expansion of management and the protection objectives of natural resources, watershed development has shifted from a traditional single task such as power generation and flood control to comprehensive utilization taking account of the ecological environment. While on the basis of the total amount of water resources availability, there are often contradictions among the development goals in the basin due to unreasonable utilization [1] such as deterioration of the ecological environment caused by the increase of irrigation water usage and large power loss as a result of large discharge in the flooding season and insufficient water after the flooding season.

The proper operation of cascade reservoirs is a multi-stage, nonlinear, high-dimensional and strictly controlled optimization process pertinent to the planning and management of water resources [2–5]. From the optimization direction, the optimization methods could be categorized as gradient search, direct search and meta-heuristic search [6]. From the development process, algorithms utilized to achieve optimal scheme of reservoir operation could be divided into classic methods and intelligence optimization algorithms. Traditionally, many researchers have adopted classical techniques such as linear programming, nonlinear programming and dynamic programming to deal with the MOPs, either a weighted or a constrained approach, without considering all objectives simultaneously [7–9]. The classic methods are relatively simple but are limited by the possibilities such as not attaining global optima, convergence to local optima and being hampered by dimensionality. Natural phenomena drive the development of the latter intelligence optimization methods. One of the strengths of intelligence optimization algorithms is their convergence to be virtually global optimal for any well-defined optimization problem [10–12]. In recent years, many intelligent optimization algorithms and improved algorithms, such as Evolutionary Algorithms (EA), Artificial Neutral Network (ANN), Non-Dominated Sorting Genetic Algorithm-II (NSGA-II), Particle Swarm Optimization (PSO) and Cuckoo Search (CS), have been proved to be useful in the multi-reservoir systems optimization for maximizing benefits of multi-objectives [13–18]. This study employs improved NSGA-II to solve the MOPs of reservoir optimal dispatching.

As the MOPs usually contain multiple incommensurable and competing objectives, it is impossible to find a single optimal solution to satisfy all the targets. Instead, the solution exists in the form of alternative trade-offs, known as the Pareto optimal solutions [19]. There is a contradiction between objectives of every Pareto optimal solution, identified as universality and particularity. To better understand the evolution process of Pareto optimality and choose the optimum, it is necessary to explore the competition mechanism among multiple objectives and the interactions between two or three objectives.

Currently, most of studies on the competition mechanism between multiple objectives always focus on the objectives of several typical schemes which are unavoidably subjective, while few studies address the competition rules among multi-objectives (≥3) of long series to reflect the statistical law of each objective [20–22]. Copulas [23] are known as an effective means for describing the dependence between random variables, thus expected to be suitable for studying competition or dependence among multi-objectives. Recently, different copulas have been employed for the multivariate analysis of spatiotemporal change in probabilistic forecasting of seasonal droughts [24], multivariate real-time droughts assessment [25], joint return periods of precipitation and temperature extremes [26], flood frequency analysis [27,28], risk analysis [29], energy environmental optimization [30–32], stochastic hydrological simulation [33] and so on, while its application in the research field of multi-objective competition relationship has not yet emerged. A key feature of copula is to characterize the dependency structure of two or more variables, either cross-correlation or auto-correlation [34], making it a promising method for the multi-objective questions. In this study, copula function is chosen to construct joint sequence values of two targets. This is the first time that copula is applied to explore multi-objective competition mechanism whereby the overall impact of one objective on the other two objectives can be evaluated.

The Heihe River Basin, as the second largest inland river basin of China, is the study area. The primary goals of this study are: (1) to analyze the three-dimensional and two-dimensional spatial distributions of the Pareto set obtained on account of reservoir dispatching obtained in the preliminary work; (2) to establish a formula for quantitatively describing the relationship between two objectives and to inform the law of water use; and (3) to explore the multi-objective competition mechanism for the overall impact of one objective on the other two objectives with the copula function constructing the joint sequence of two targets. The goal of this study is to reveal the multi-objective competitive mechanism and alleviate the competitive nature of water usage. It also provides guidance for the unified allocation of water resources and provides insights for other multi-objective problems.

2. Study Area and Data

As the second largest inland river basin of China, Heihe River Basin is situated within 98° E~102° E and 37.5° N~42.4° N, with an area of about 134,000 km^2. With an average annual precipitation of 400 mm and an average annual potential evaporation of 1600 mm, the basin lays in the interior of the Eurasian continent and characterized by arid hydrological features. The predominant land use types are desert and grass lands, accounting for roughly 60% and 25% of the total area, respectively [35]. Under water and ecological stresses, the Heihe River Basin suffers water-table decline, terminal lakes dryness, grassland degeneration as well as widespread desertification largely owing to the combined impacts of climate change and human activities [36,37]. Given its key role in water resources planning and management in northwestern China, the Heihe River Basin has long been a focus of studies on inland rivers in arid regions and it is therefore selected for this case study.

The main stream of Heihe River is about 928 km long from the birthplace of the Qilian Mountains to the tail of Juyan Lake. It is divided into upstream, middle and downstream areas by the Yingluo Gorge and the Zhengyi Gorge. The upstream is the main water production area. The middle is the main irrigation water area. The downstream is the main ecological water consumption area (as shown in Figure 1). The first control project in the Heihe river basin is Huangzangsi hydro-junction, covering an area of 7648 km^2 and controlling almost 80% incoming water in the upper reaches of the Heihe River (Figure 1). Huangzangsi is a within-year reservoir which has 4.06×10^8 m^3 total storage and 3.34×10^8 m^3 regulating storage, with 6.02 MW guaranteed output and 49 MW installed capacity. Huangzangsi Reservoir and seven run-off hydropower stations, which are Baopinghe, Sandaowan, Erlongshan, Dagushan, Xiaogushan, Longshou-II and Longshou-I, combining with the Zhengyixia Reservoir at the end of the middle reaches, to constitute the structure of cascade reservoirs that have '2 reservoirs 7 stations.'

In the previous work, based on the long series of monthly run-off data from July 1957 to June 2014, the model of cascade reservoirs dispatching in the Heihe River Basin has been established and solved, which will be the basis of this study.

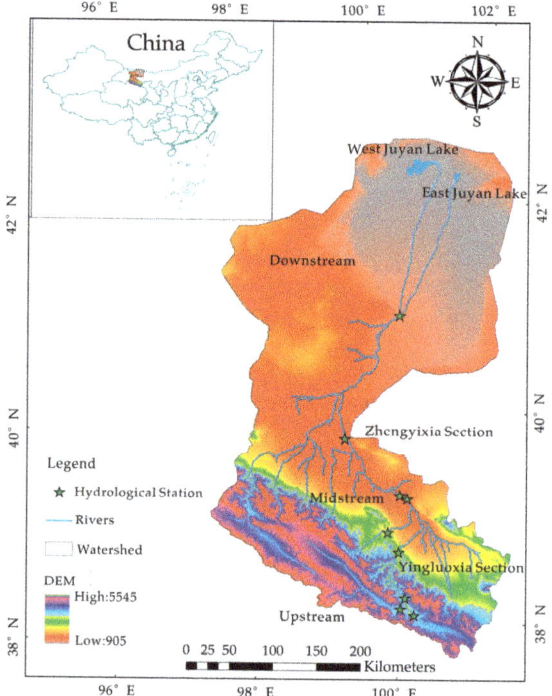

Figure 1. Location and sections distribution of the Heihe River Basin.

3. Research Methodology

3.1. Description of Multi-Objective Model of the Heihe Cascade Reservoirs Operation

Objective Function

The comprehensive utilization of water resources in the main stream of Heihe River is mainly embodied by three aspects: power generation in the upper reaches, irrigation in the middle reaches and ecological water use in the lower reaches. Consequently, this configuration has three objectives, namely minimum irrigation water shortage, minimum ecological water shortage and maximum generation capacity of cascade hydropower stations. The allocation of water resources in the Heihe River Basin requires that the 'electricity regulation' complies with the 'water regulation,' that is, the irrigation water and the ecological water need to be given priority while taking into account power generation requirements. With the decision variables of the final water level of the time interval of Huangzangsi Reservoir and Zhengyixia Reservoir, the objective functions are as follows:

Objective 1: Minimum Irrigation Water Shortage

$$\min f_1 = \sum_{t=1}^{T} \sum_{i=1}^{I} \alpha(t,i)\left[Q_{irr}^{d}(t,i) - Q_{irr}^{s}(t,i)\right]\Delta t \tag{1}$$

$$\alpha(t,i) = \begin{cases} 0 \; Q_{irr}^{d}(t,i) \le Q_{irr}^{s}(t,i) \\ 1 \; Q_{irr}^{d}(t,i) > Q_{irr}^{s}(t,i) \end{cases} \tag{2}$$

Objective 2: Minimum Ecological Water Shortage

$$\min f_2 = \sum_{t=1}^{T} \beta(t)\left[Q_{eco}^{d}(t) - Q_{eco}^{s}(t)\right]\Delta t \tag{3}$$

$$\beta(t) = \begin{cases} 0 \; Q_{eco}^{d}(t) \le Q_{eco}^{s}(t) \\ 1 \; Q_{eco}^{d}(t) > Q_{eco}^{s}(t) \end{cases} \tag{4}$$

Objective 3: Maximum generation capacity

$$\max f_3 = \sum_{t=1}^{T} \sum_{j=1}^{J} K_j Q_{ge}(t,j) H(t,j) \Delta t \tag{5}$$

Subject to the following constraints:

$$V(t,j) = V(t-1,j) + [Q_{\text{in}}(t,j) - Q_{\text{out}}(t,j)]\Delta t \tag{6}$$

$$Z_l(j) \le Z(t,j) \le Z_h(j) \tag{7}$$

$$Q_{\min}(j) \le Q_{\text{out}}(t,j) \le Q_{\max}(j) \tag{8}$$

$$0.6 \times N_g(j) \le N(t,j) \le N_c(j) \tag{9}$$

All variables need to satisfy non-negative constraints where, f_1 (10^4 m^3) is irrigation water shortage; f_2 (10^4 m^3) denotes ecological water shortage; f_3 (10^8 kW·h) indicates generation capacity of cascade hydropower stations; t is time interval number and T is total number of time intervals, $T = 684$; i is river number and I is total river reach number, $I = 5$; $Q_{irr}^{d}(t,i)$ and $Q_{irr}^{s}(t,i)$ indicates irrigation water demand and water supply flow of time t and river i respectively, m^3/s; $\alpha(t,i)$ denotes the irrigation coefficient of time t and river I; Δt is unit time, $\Delta t = 2.63 \times 10^6$ s; $Q_{eco}^{d}(t)$ and $Q_{eco}^{s}(t)$ represents the ecological water demand and water supply flow of time t respectively; $\beta(t)$ is the ecological coefficient in time t; j *is* hydropower station number and J is total number of hydropower stations, $J = 9$; K_j is comprehensive output coefficient of station j; $Q_{ge}(t,j)$ indicates generation flow in time t of station j; $H(t,j)$ (m) denotes the net water head in time t of station j; $V(t,j)$ indicates the terminal storage capacity of time t and station j; $Q_{\text{in}}(t,j)$ and $Q_{\text{out}}(t,j)$ denote the input and outflow. $Z_l(j)$ and $Z_h(j)$ describe the lowest water level and the highest water level for power generation in Reservoir j; $Z(t,j)$ is the final reservoir level of time t and station j; $Q_{\min}(j)$ and $Q_{\max}(j)$ indicates the minimum discharge and maximum safe discharge of station j; $N_g(j)$ and $N_c(j)$ are the guarantee output and installed capacity respectively of station j; $N(t,j)$ is the output of time t and station j.

The constraints include: reservoir water balance constraint; reservoir water level constraint; reservoir discharge; hydropower station output; boundary initial conditions; and non-negative variables.

3.2. *ICGC-NSGA-II Algorithm*

NSGA-II is recognized as one of the most effective multi-objective optimization algorithms. By non-dominated sorting of multi-objective problems, NSGA-II replaces the traditional multi-objective methods in transforming objective functions (such as weighted or constrained transformation). Its inherent parallelism makes it possible to search multi-objective non-inferior solution sets simultaneously and to deal with non-inferior frontier irregular optimization problems [7,38,39]. Moreover, it can better maintain the diversity of the population and has a strong robustness. However, multi-objective reservoir dispatching is a multi-stage decision-making problem. The value of decision variables in the previous stage usually affects the scope of the feasible region in the next stage. When evolutionary algorithm is adopted to solve the problem, the feasible region takes a small proportion in the search space and the efficiency of the evolutionary algorithm is very low. Reducing the proportion of infeasible areas in search area is thus one of the most effective methods for improving the efficiency of operation. In this study, ICGC-NSGA-II is used to solve the reservoir dispatching model and the specific procedures can be found in Reference [21].

3.3. Normalization of Objective Sequence

In this study, the three target values are normalized. To eliminate the difference of dimension of each objective and make them comparable, each object is normalized according to the following formula:

$$y_{ij} = \left[x_{ij} - \min\left(x_{ij}\right)\right] / \left[\max\left(x_{ij}\right) - \min\left(x_{ij}\right)\right] \tag{10}$$

where, x_{ij} and y_{ij} are target eigenvalue and the normalized value of the non-inferior solution $j(j = 1,2,\ldots,n)$ of the object $i(i = 1,2,\ldots,m)$ respectively; Δx_{ij} is the difference between the target eigenvalue and the moderate value.

Among the three objectives of this study, the average irrigation water shortage (Obj-1) and ecological water shortage (Obj-2) are the minimizing objectives. Conversely, the multi-year average generation capacity (Obj-3) belongs to the maximizing objective. The optimization directions of objectives are different. In order to construct the joint sequence values of two targets, the optimization direction of normalized values needs to be consistent. The normalized value of Obj-1 and Obj-2 were subtracted by 1. A new normalized matrix is generated by replacing the original normalized value with the difference.

3.4. Copulas Theory

The copula was originally introduced by Sklar et al. [40] as a useful method to derive joint distributions with marginals. Its design allows it to deal with non-normal distributions. This method is also likely to be beneficial when the marginals can be definitely stipulated but the joint distribution is not straightforward to construct. The name 'copula' basically comes from the word 'couple' to emphasize a manner whereby a joint distribution function and the marginals can be combined [23]. The copula is a multidimensional joint probability distribution function with a uniform distribution in [0,1] interval. It can connect multiple random variables $F(x_1,x_2,\ldots,x_n)$ with their respective marginal distribution functions $F_1(x_1),F_2(x_2),\ldots,F_n(x_n)$ for the joint distribution function. Therefore, copula is often referred to as 'connection functions' or 'dependent functions.' Among them, the marginal distribution function describes the distribution of a single variable, while the copula function indicates the joint distribution of multi-dimensional composite variables [41].

The 'Sklar theorem' defines that: Assuming F is an n-dimensional distribution function whose marginals are $(F_1, F_2, \ldots, F_n)$ of a random vector $(X_1, X_2, \ldots, X_n)$, there exists an n-dimensional copula function C, which satisfies the following formula for arbitrary $x \in R^n$:

$$F(x_1,x_2,\ldots,x_n) = C[F_1(x_1),F_2(x_2),\ldots,F_n(x_n)] = C(u_1,u_2,\ldots,u_n) \tag{11}$$

If the marginal distributions $F_1, F_2, \ldots, F_n$ are continuous, function C is uniquely determined. Conversely, if C is an n-dimensional copula function and $F_1, F_2, \ldots, F_n$ are a set of univariate distribution functions, the function F defined by Equation (11) is an n-dimensional distribution function having marginal distribution as $F_1, F_2, \ldots, F_n$.

Copula functions could be divided into four categories as a whole: Archimedean Copula, Extreme Copula, Elliptical Copulas and other hybrid families of copulas. In this study, two-dimensional copula of Clayton Copula, Frank Copula and Gumbel Copula in the Archimedean Copula function family and Gaussian Copula and Student Copula in the elliptic population are selected.

Copula is able to construct multi-dimensional joint probability distribution function by arbitrary marginal distribution and correlation structure [42]. It is flexible in forms and does not require all variables to obey the same scatter. In recent years, copula is often used to analyze the joint recurrence period of multi-variables and the frequency of combined events. Copula function has natural advantages in constructing joint distribution of two variables. In this study, copula is used to construct joint sequence values of two targets and applied to the study of multi-objective competition mechanism for the first time. Procedures of constructing joint sequence values of two targets are as follows:

1. Dependence measurement. Before constructing multivariate joint distribution, it is necessary to measure the correlation between different random variables according to the correlation index.
2. Marginal distribution fitting. The marginal distribution of each single variable should be fitted to find the appropriate distribution type. Since all the parametric methods in this study have not passed the test, the non-parametric method is adopted here. In this paper, a non-parametric empirical frequency determination method based on the Gringorten formula [43,44] is introduced.
3. Parameter estimation of copula function. In this paper, the maximum likelihood estimation method [45] is used to estimate parameters.
4. Goodness of fit evaluation. The goodness-of-fit evaluation is an important way of comparing and analyzing the goodness-of-fit evaluation indices of different types of copula functions, so as to optimize the most suitable distribution of copula functions.

The AIC (Akaike Information Criterion) method is proposed by Akaike in 1974 [46] from the perspective of information theory. It is a goodness-of-fit evaluation criterion based on the Kullback-Leibler information metric, which contains two factors: the deviation between the empirical points and theoretical copula functions as well as the error fluctuation caused by the number of parameters of copula functions. Taking the joint copula function of two variables as an example, the concrete formulas of the AIC information criterion method are as follows:

$$MSE = \frac{1}{n}\sum_{i=1}^{n}(F_n(x_i, y_i) - C(u_i, v_i))^2 \tag{12}$$

$$AIC = n\ln(MSE) + 2m \tag{13}$$

where, $F_n(x_i, y_i)$, $C(u_i, v_i)$ are the empirical and theoretical frequencies of the joint copula function with two variables are expressed, respectively. MSE is the mean square error and m is the number of parameters of the model.

Among them, the smaller the value of AIC, the better the fitting of the copula function. The AIC information criterion is applicable to the comparison and optimization of copula functions with a different number of model parameters.

5. Computation of Joint Distribution Sequence Values. In order to find a new sequence that reflects the overall characteristics of the two variables, it needs to inverse the frequency sequence obtained above and then get the joint sequence value of the two variables. In this study, the inverse function, that is, NORMINV for the normal cumulative distribution function in the MATLAB software is used to derive the sequence values of the joint probability distribution of the copula function.

In this paper, with the Pareto set output from the multi-objective optimal operation model as the input, copula is employed to study the multi-objective competition mechanism. The main research process is as follows (Figure 2):

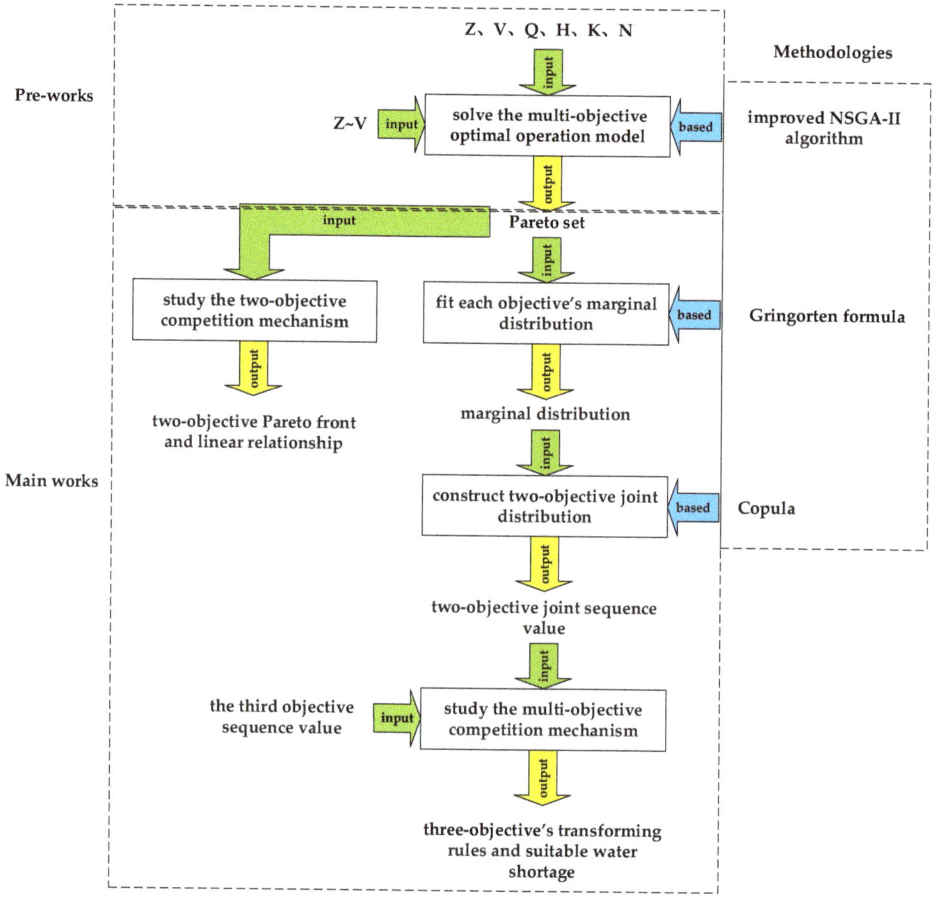

Figure 2. The main research framework.

4. Results and Discussions

4.1. Optimal Results of the ICGC-NSGA-II

In the main stream of the Heihe River, Huangzangsi Reservoir, seven run-off hydropower stations of Baopinghe, Sandaowan, Erlongshan, Dagushan, Xiaogushan, Longshou-II and Longshou-I and Zhengyixia Reservoir at the end of the middle reaches constitute the structure of '2 reservoirs 7 stations' as the cascade reservoirs' hydropower station system. With the ICGC-NSGA-II algorithm for solving the optimal dispatching multi-objective model, the decision variables are the final water level of the time interval of Huangzangsi Reservoir and Zhengyixia Reservoir. The population size is 2000 and the number of iterations is 500 generations. The results of long-term optimal scheduling are sorted according to the ascending order of annual water shortage, as shown in Table 1.

According to Table 1, the minimum and maximum annual average water shortage of irrigation are 925×10^4 m^3 and 6595×10^4 m^3; the minimum and maximum of the ecological water shortage are 3925×10^4 m^3 and 9058×10^4 m^3; and the minimum and maximum of the multi-year average power generation are 25.54×10^8 kW·h and 26.99×10^8 kW·h, respectively. The non-inferior solution set is plotted in one coordinate system and the Pareto surface maps of irrigation, ecology and power generation are obtained, as shown in Figure 3.

Table 1. Multi-objective Pareto optimal set.

No.	Annual Water Shortage in Irrigation (10^4 m^3)	Annual Water Shortage in Ecological (10^4 m^3)	Annual Generation Capacity (10^8 kW·h)
1	925	9058	25.95
2	926	9010	25.94
3	935	9000	25.96
...	...	...	...
1998	6580	4396	26.34
1999	6584	4400	26.34
2000	6595	4424	26.35

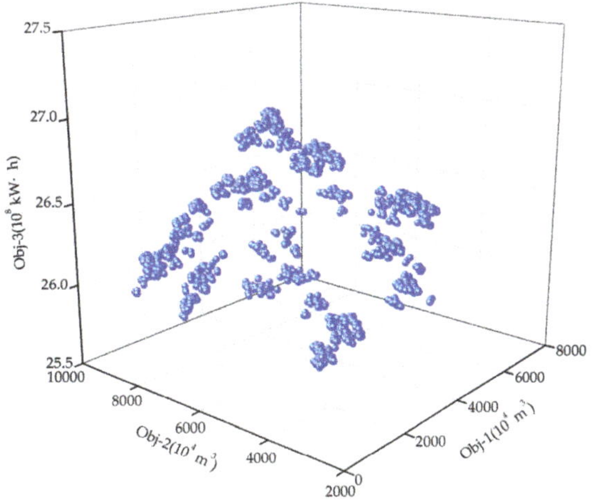

Figure 3. Multi-objective Pareto surface graph of Obj 1~3 in one coordinate system. Blue points indicates non-inferior solution.

Intuitively, the three-dimensional spatial distribution and the surface shape of the multi-objective non-inferior solution are shown in Figure 3. The non-inferior solutions appear in clusters in the Pareto surface graph (Figure 3) and are distributed over a wide range. The results show that there is little difference among non-inferior solutions in the same cluster and the optimal scheduling results converge well near the clusters to which the non-inferior solution belongs. The algorithm itself retains the diversity of non-inferior solutions. As the basic regulating rules, the Obj-1 and Obj-2 are 'the smaller the better' and the Obj-3 is the larger the better. Hence, theoretically, the optimal solution for the scheme is '925 × 10^4 m^3, 3925 × 10^4 m^3, 26.99 × 10^8 kW·h.' However, the three objectives cannot reach the optimal solution of the sample theory at the same time—an indication of mutual restrictions and influences among the three objectives. The Pareto surface of the scheme is a smooth space surface oriented to the vector ± $(-1, -1, 1)$ direction (the main vertical line of the surface is consistent with the direction of a vector, which is expressed as the surface oriented to the vector direction). It can be inferred that the set of non-inferior solutions is composed of the non-inferior solution, which is weighted by the equivalent distance from the optimal solution of the sample theory. Vector ± $(-1, -1, 1)$ is the superior direction of non-inferior solutions, consistent with the direction of the non-inferior solutions set towards the optimal solution of the sample theory, while vector ± $(1, 1, -1)$ is the inferior direction.

It is difficult to exhibit the mutual restriction among different objectives; therefore, the three-dimensional stereogram was converted into two-dimensional plane map [47] and three objectives are respectively described in terms of bubble size. The bigger the bubble, the better the objective. Thus, the multi-objective bubble diagram is generated and shown in Figures 4–6.

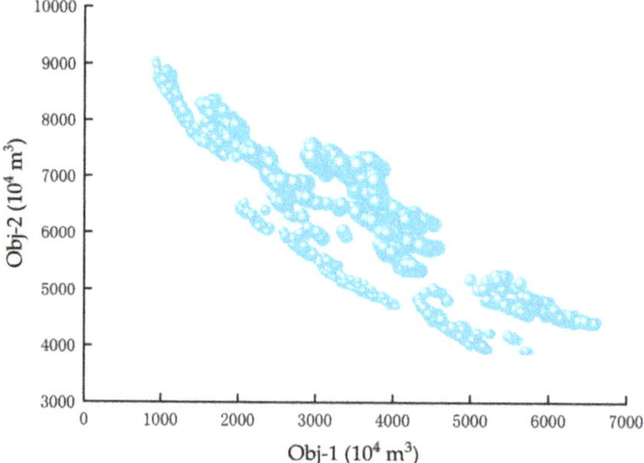

Figure 4. Bubble graph of irrigation and ecology, in which the bubble size represents the multi-year average power generation (Obj-3).

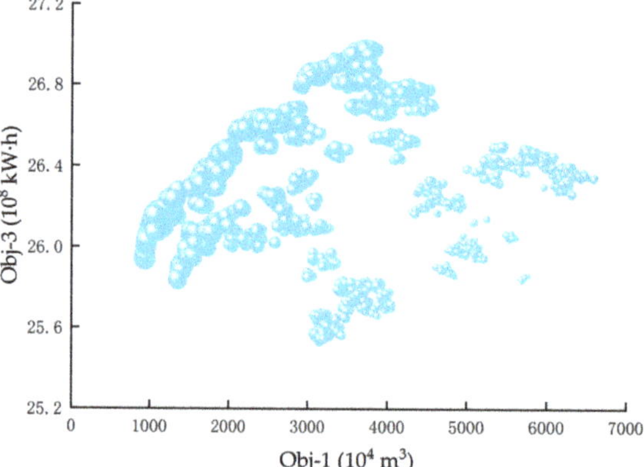

Figure 5. Bubble graph of irrigation and power generation, in which the bubble size represents the multi-year average ecological water shortage (Obj-2).

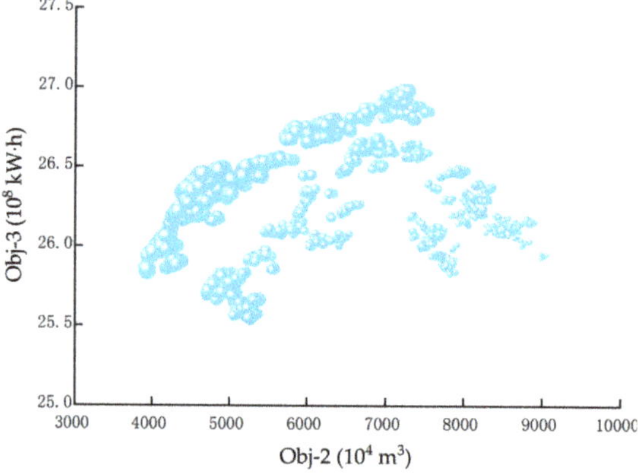

Figure 6. Bubble graph of ecology and power generation, in which the bubble size represents the multi-year average irrigation water shortage (Obj-1).

The bubble diagram shows an approximate linear distribution with layers (Figures 4–6). Taking the bubble graph of irrigation and ecology (Figure 4) as an example, the majority distribution of non-inferior solutions spreads wider. The stratification phenomenon and the wide distribution are supportive of the convergence and diversity of the optimal dispatching results. The approximate linear distribution indicates that there exists a distinct growth-decline phenomenon and a direct water competition relationship between the irrigation water shortage and the ecological water shortage. It is notable that the closer to the intersection of the two coordinate axes, the smaller the bubble; while the farther away from the intersection, the larger the bubble. Within the same layer, the bubble size is equivalent. According to the basic regulating rules, it is evident that the closer the equivalent weighted distance from the intersection, the better the non-inferior solution and the larger the air bubbles. The distribution of bubble size illustrates the competitive relationship between power generation with irrigation and ecological water, that is, the smaller the water shortage of irrigation and ecology, the less power generation capacity. The contrary indicates more power generation.

4.2. Analysis of Two-Objective Competition Mechanism

As a way of exploring the competition mechanism and quantitatively describe the transformation law among objectives, the Pareto frontier is fitted to obtain the quantitative transformation formula between any two objectives. According to the previous analysis, the three-objective Pareto solution set is fully optimized and most of the solutions are high-quality feasible solutions close to the optimum one. On the premise of sufficient optimization, there is a macro-rule of 'one falls another rises.' By sorting the normalized values of one objective and drawing two-dimensional scatter plots of the normalized indices of the other two, several solutions are intercepted using the normalized values from small to large. It can be found that these solutions are exactly the Pareto frontier of the other two objectives. In other words, when the normalized value of one objective is small (that is, the solution is inferior), then the other two targets show the strongest regularity and optimum. Additionally, the number of interception solutions for sorting targets depends on the coverage of the Pareto frontier of the other two targets.

As shown in Figure 7, the two-dimensional scatter plots of Obj-1 and Obj-2 normalization indices, the non-inferior solutions are sorted in descending order according to the normalized value of Obj-3. After 700 solutions are intercepted, the Pareto frontier of Obj-1&2 is obtained as the orange points (Figure 7). A mutation point (0.7968, 0.4905) is found in the red dot (Figure 7).

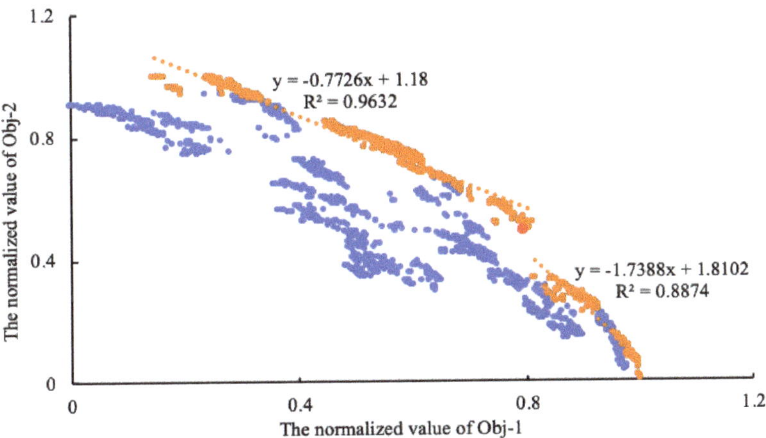

Figure 7. The two-dimensional scatter plots of Obj-1 and Obj-2's normalized value. All the points indicate Pareto set, and among them, the orange points represent the Pareto frontier.

The scatter plots are drawn by selecting two sequences before and after the mutation points in the Pareto frontier (Figure 7). Then, the piecewise linear function relationship of the two series is obtained, as shown in Figure 7. The correlation coefficients of the two series are 0.981 and 0.942, respectively.

From the above formula, it can be seen that both the slopes of the two piecewise functions are less than zero and the normalized values of Obj-1 and Obj-2 are inversely correlated.

In front of the mutation point, the independent variable fluctuates within the range of [0,0.7968] and the normalized value of Obj-2 decreases by 0.7726 for each increase of Obj-1. Following the mutation point, the range changes to [0.7968,1] and the normalized value of Obj-2 decreases by 1.7388 for each increase of Obj-1. Converting to their own respective target values of the scheme, when the independent variable is between 925~2078 $\times$ 10^4 m^3, the average annual ecological water shortage will increase by 9.0553 $\times$ 10^4 m^3 for each reduction of 1 $\times$ 10^4 m^3 of the average annual irrigation water shortage. Yet, when the independent variable is between 2078~6595 $\times$ 10^4 m^3, the average annual ecological water shortage will increase by 0.6825 $\times$ 10^4 m^3 for each reduction of 1 $\times$ 10^4 m^3 of the average annual irrigation water shortage. Conclusively, in the interval where the irrigation water shortage is large, the increase of the ecological water shortage is slow while reducing the irrigation water shortage. In the interval where the irrigation water shortage is small, continuing to reduce the irrigation water shortage will lead to a significant increase in the ecological water shortage.

The two-dimensional scatter plot of the normalized indices of 'Obj-1 and Obj-3' and 'Obj-2 and Obj-3' is shown in Figures 8 and 9 in the way similar to Figure 7, with correlation coefficients of 0.984 and 0.933, respectively.

From Figure 8, it can be seen that the slope of the trend line is negative and the normalized values of Obj-1 and Obj-3 are inversely correlated. The normalized value of Obj-3 decreases by 1.3009 units for each additional unit of Obj-1 normalized value. An indication is that for each reduction of the average annual irrigation water shortage by 100 $\times$ 10^4 m^3, the average annual power generation decreases by 0.03349 $\times$ 10^8 kW·h. The average annual irrigation water shortage is positively correlated with the average annual power generation.

From Figure 9, the slope is also negative and the normalized values of Obj-2 are inversely correlated with Obj-3. The normalized value of Obj-3 decreases by 1.0687 units for each additional unit of Obj-2 normalized value. The inference of the results is that: for each reduction of the average annual ecological water shortage by 100 $\times$ 10^4 m^3, the average annual power generation decreases by 0.03443 $\times$ 10^8 kW·h. There is a positive correlation between the average annual irrigation water shortage and the average annual power generation.

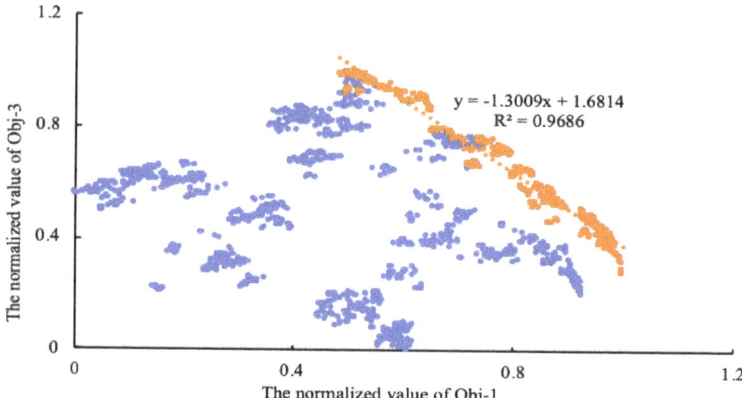

Figure 8. The two-dimensional scatter plots of Obj-1 and Obj-3's normalized value. All the points indicate Pareto set, and among them, the orange points represent the Pareto frontier.

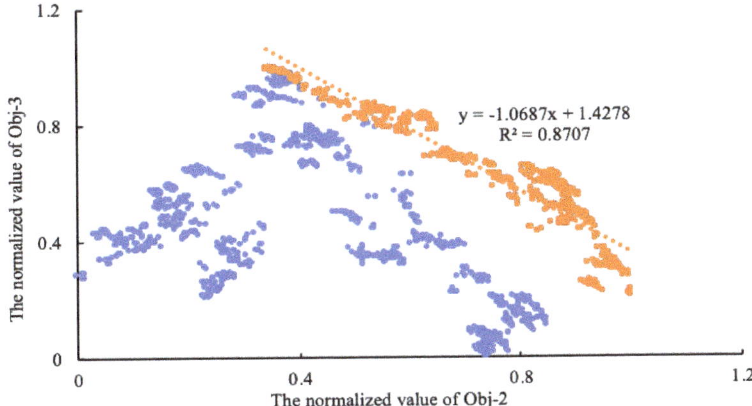

Figure 9. The two-dimensional scatter plots of Obj-2 and Obj-3's normalized value. All the points indicate Pareto set, and among them, the orange points represent the Pareto frontier.

4.3. Analysis of Three-Objective Competition Mechanism

Considering that the objectives of the Heihe River Basin are integrally and mutually restrictive, it is virtually impossible to reveal the law of the whole water resources system only by studying the relationship between the two objectives. This section synthesizes the sequence values of any two targets into a new sequence that represent the characteristics and information of two sub-sequences and analyses impact of the change of one target on the whole of the other targets in the Heihe River Basin. Through these analyses, the competition mechanism among the three objectives is explored.

Dependence between hydrological series is the premise of constructing the joint distribution using multivariate copula. The previous study indicates that the correlation coefficient between any two objectives is within 0.933~0.984. Such a high correlation supports the construction of the joint copula function of two variables.

The marginal distribution of each target is constructed with the non-parametric method of Gringorten and the empirical frequency estimates of each target sequence are obtained. The maximum likelihood method is used to get the parameter estimates of five joint copula functions between the two target sequence values and the corresponding cumulative distribution function values are calculated. The goodness of fit of five copula functions is evaluated with the AIC method which are best matching with existing hydrological sequences. The AIC evaluation indices of five combined copula functions under different objective combinations are obtained as shown in Table 2.

Table 2. Five copula functions' Akaike Information Criterion (AIC) evaluating value.

Objective Combination	Clayton Copula	Frank Copula	Gumbel Copula	Gaussian Copula	Student Copula
Obj-1&2	-8635.43	−16,051.64	-8635.43	−16,435.24	−16,437.86
Obj-1&3	−10,140.03	−10,629.38	−10,140.03	−10,671.15	−10,669.85
Obj-2&3	−10,075.94	−11,292.73	−10,075.93	−11,282.21	−11,281.74

According to the AIC information criterion, the smaller the value of AIC evaluation index, the better the fit of representative copula function. As can be seen from Table 2 above, the minimum AIC evaluation index of Obj-1&2 is −16,437.86 with the Student Copula function. Similarly, the minimum in Obj-1&3 and Obj-2&3 are −10,671.15 (Gaussian Copula) and −11,292.73 (Frank Copula) respectively. As a result, the three two-variable-copula functions are selected as the optimal bivariate copula joint distribution for combining sequences.

To derive the sequence value of joint distribution, the joint sequence value of the optimal copula function under three combinations is obtained. The joint sequence value covers the characteristics and the information of the two objective sequences that represent the overall level. The scatter plot of the Obj-1 normalized value and the Frank Copula joint sequence value of Obj-2&3 is illustrated in Figure 10, as well as the 'Obj-2 + Obj-1&3(Gaussian Copula)' and 'Obj-3 + Obj-1&2(Student Copula)' showed in Figures 11 and 12, respectively.

Comparing Figures 10–12, Figure 12 appears the most discrete, which indicates that the regularity of the impact of Obj-3 on Obj-1&2 in the Heihe River Basin is the worst. It is known that power generation does not consume but rather utilizes surface water resources, so the competition between power generation and the other objectives is minimal. Moreover, the slope is the smallest, indicating that the change of power generation has the least impact on Obj-1&2.

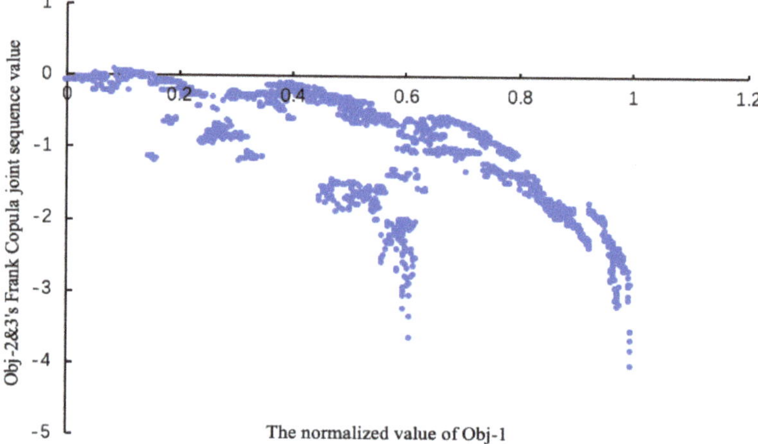

Figure 10. The normalized value of Obj-1 and Obj-2&3's joint series scatter diagram.

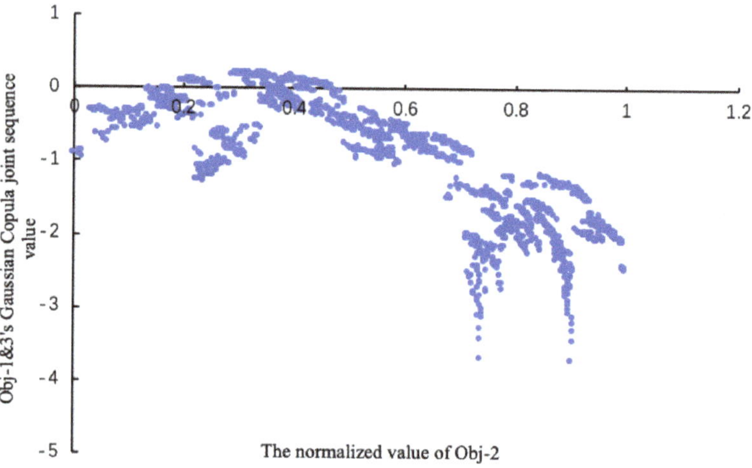

Figure 11. The normalized value of Obj-2 and Obj-1&3's joint series scatter diagram.

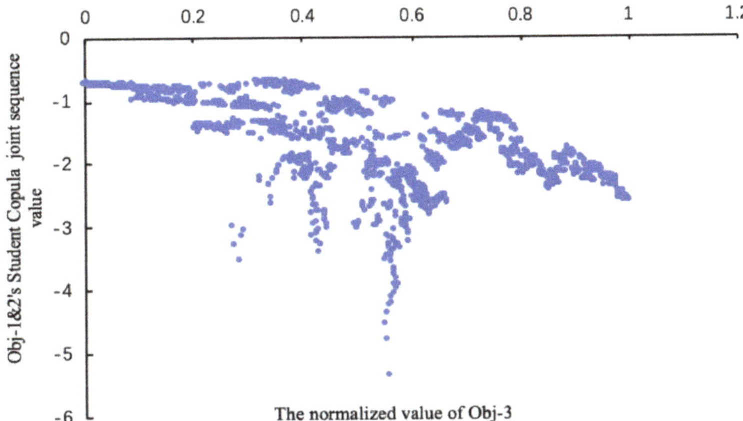

Figure 12. The normalized value of Obj-3 and Obj-1&2's joint series scatter diagram.

The increase of Obj-1 normalization index will lead to the decrease of combined sequence value of Obj-2&3, which indicates that when Obj-1 tends to be more optimized, the Obj-2&3 will become worse as a whole. The slope of the Obj-1 normalized value in scatter plot ahead of 0.9 is milder than that after 0.9, which denotes that when irrigation water shortage is greater than 1492×10^4 m^3, with the decrease of irrigation water shortage, the overall impact on ecological water and power generation is smaller and the cost of the optimal irrigation water shortage is lower. When Obj-1 is more satisfied, that is, when irrigation water shortage is less than 1492×10^4 m^3, with the decrease of irrigation water shortage, the overall impact on ecological water and power generation is greater and the cost of the optimal irrigation water shortage is higher. Therefore, it is recommended that the average annual water shortage for irrigation should be about 1492×10^4 m^3.

The plot of the Obj-1 normalization index is scattered before 0.6 and clustered after 0.6. It illustrates that when the irrigation water shortage is greater than 3193×10^4 m^3, the change of irrigation water has little effect on the ecological water and power generation integrally; after 0.6, when the irrigation water shortage is less than 3193×10^4 m^3, the irrigation water exhibits a strong influence on other objectives.

The increase of the Obj-2 normalization index will lead to the decrease of the combined sequence value of Obj-1&3, which indicates that when the Obj-2 is more optimized, the Obj-1&3 may become worse as a whole. The slope of the Obj-2 normalization index in the scatter plot before 0.8 is milder than that after 0.8, which denotes that, in the stage of lower ecological satisfaction, when ecological water shortage is greater than 4951×10^4 m^3, with the decrease of ecological water shortage, the overall impact of irrigation water and power generation is smaller and the cost of optimizing the ecological water shortage is lower. Otherwise, when ecological water shortage is less than 4951×10^4 m^3, with the decrease of ecological water shortage, the overall impact on irrigation water and power generation is greater and the cost of optimizing ecology water shortage is higher. It is recommended that the average annual water shortage in the ecological process should be around 4951×10^4 m^3.

The increase of normalized index of Obj-3 will lead to a decrease of the combined sequence value of Obj-1&2, which indicates that when the Obj-3 tends to be more optimized, the Obj-1&2 will become worse as a whole. The scatter plots before 0.65 are discrete and after 0.65 are concentrated, showing that when the average annual power generation is less than 26.48×10^8 kW h, the change of power generation has little effect on the irrigation and ecological water as a whole; on the contrary, the impact is stronger.

5. Conclusions

This study takes the multi-objective joint optimal dispatch of cascade reservoirs in the Heihe River Basin as a study object. Based on the ICGC-NSGA-II algorithm to solve this model, the Pareto non-inferior solution set is obtained. The competition mechanism among two objectives is quantitatively described by statistical means and with the copula function constructing the joint sequence of two targets, the three objectives' competition mechanism is explored too. This study expects to provide a foundation for selective preference of the Pareto set and a new idea for multi-objective study. The main conclusions are summarized as follows:

1. The three-dimensional and two-dimensional spatial distributions of the Pareto solutions prove that there are mutual restrictions and influences among the three objectives. In order to avoid the disadvantage of choosing only a limited number of representative solutions and being too arbitrary, the long series of non-inferior solutions obtained are adopted to study the competition mechanism in this study. On the premise of sufficient optimization, there is a macro-rule of 'one falls another rises.' When one objective solution is inferior, then the other two targets show the strongest regularity and optimum.
2. In the analysis of the two-objective competition mechanism, the functional formulas between the sequences of two objects are given, which can quantitatively describe the relationship and interactions. It was found that when the irrigation water shortage was large, with it decreasing, the ecological water shortage increased slowly, which indicates that the two are inversely correlated. In addition, there is a positive correlation between the multi-year average irrigation water shortage and the average power generation, as there is between ecological water shortage and power generation.
3. This study first applied the two-variable joint copula function to the study of the multi-objective competition mechanism. Based on the advantage that copula function cannot produce information distortion in the process of connecting the marginal distribution of two sub-sequences, a new sequence containing the comprehensive information of the two targets is generated by using the joint copula function of two variables to combine the sequence values of any two objectives and the competition mechanism between the remaining target sequence and the joint sequence of two targets is studied. A new way is provided for studying the influence of a single sequence on the compound sequence of two sequences.
4. The three-objective competition mechanism infers that the competition between power generation and other objectives is the least and the change of power generation has the least influence on the other two as a whole. Specifically, the recommended annual average water shortage for irrigation is about 1492×10^4 m^3. When it is less than this value, with decreasing irrigation water shortage, the overall impact of ecological water and power generation is greater. Only when the irrigation water shortage is less than 3193×10^4 m^3, will there be a strong impact on other objectives. Additionally, the average annual ecological water shortage is about $4951 \times 10^4 m^3$, when it is less than this value, the overall impact of the irrigation water and power generation will be greater as the ecological water shortage decreases. After the average generation capacity has been more than 26.48×10^8 kW h for many years, the objective of power generation has a strong influence on the other targets.

To summarize, the copula function could combine the marginal distribution of any two sequences and construct a new joint sequence and all the information of the sub-sequence is contained, so there is no information distortion during the combination process. It is an effective tool for quantitatively studying the multi-objective competition mechanism. At present, the research on multi-objective competition mechanisms is in the preliminary stage and the methods adopted in this paper enrich this field.

Author Contributions: Conceptualization, software, data curation and writing—original draft preparation, M.Z.; methodology, L.W.; validation, H.W.; formal analysis, G.L.; resources, S.H.; writing—review and editing, S.L.; supervision, project administration and funding acquisition, Q.H.

Funding: This study was jointly funded by the National Key Research and Development Program of China (grant number 2017YFC0405900), the National Natural Science Foundation of China (grant number 51709221), the Planning Project of Science and Technology of Water Resources of Shaanxi (grant numbers 2015slkj-27 and 2017slkj-19), the China Scholarship Council (grant number 201608610170) and the Doctorate Innovation Funding of Xi'an University of Technology (grant number 310-252071712).

Conflicts of Interest: The authors declare no conflict of interest.

References

1. Asvini, M.S.; Amudha, T. Design and development of bio-inspired framework for reservoir operation optimization. *Adv. Water Resour.* **2017**, *110*, 193–202. [CrossRef]
2. Guan, J.; Kentel, E.; Aral, M.M. Genetic Algorithm for Constrained Optimization Models and Its Application in Groundwater Resources Management. *J. Water Resour. Plan. Manag.* **2008**, *134*, 64–72. [CrossRef]
3. Mathur, Y.P.; Nikam, S.J. Optimal Reservoir Operation Policies Using Genetic Algorithm. *Int. J. Eng. Technol.* **2009**, *1*, 184–187. [CrossRef]
4. Rodrigues, D.; Pereira, L.A.M.; Nakamura, R.Y.M.; Costa, K.A.P.; Yang, X.-S.; Souza, A.N.; Papa, J.P. A wrapper approach for feature selection and optimum-path forest based on bat algorithm. *Expert Syst. Appl.* **2014**, *41*, 2250–2258. [CrossRef]
5. Jiang, Z.; Ji, C.; Qin, H.; Feng, Z. Multi-stage progressive optimality algorithm and its application in energy storage operation chart optimization of cascade reservoirs. *Energy* **2018**, *148*, 309–323. [CrossRef]
6. Yang, T.; Hsu, K.; Duan, Q.; Sorooshian, S.; Wang, C. *Method to Estimate Optimal Parameters*; Springer Nature: Basingstoke, UK, 2018; pp. 1–39.
7. Reddy, M.J.; Kumar, D.N. Optimal reservoir operation using multi-objective evolutionary algorithm. *Water Resour. Manag.* **2006**, *20*, 861–878.
8. Tauxe, G.W.; Inman, R.R.; Mades, D.M. Multiobjective dynamic programing with application to a reservoir. *Water Resour. Res.* **1979**, *15*, 1403–1408. [CrossRef]
9. Liang, Q.; Johnson, L.E.; Yu, Y.S. A Comparison of two methods for multiobjective optimization for reservoir operation 1. *JAWRA J. Am. Water Resour. Assoc.* **1996**, *32*, 333–340. [CrossRef]
10. Momtahen, S.; Dariane, A.B. Direct Search Approaches Using Genetic Algorithms for Optimization of Water Reservoir Operating Policies. *J. Water Resour. Plan. Manag.* **2007**, *133*, 202–209. [CrossRef]
11. Yuan, X.; Zhang, Y.; Yuan, Y. Improved self-adaptive chaotic genetic algorithm for hydrogenation scheduling. *J. Water Resour. Plan. Manag.* **2008**, *134*, 319–325. [CrossRef]
12. Yang, X. A new metaheuristic bat-inspired algorithm. In *Nature Inspired Cooperative Strategies for Optimization*; Studies in Computational Intelligence, NISCO; Springer: Berlin/Heidelberg, Germany, 2010; Volume 284.
13. Yang, T.; Gao, X.; Sellars, S.L.; Sorooshian, S. Improving the multi-objective evolutionary optimization algorithm for hydropower reservoir operations in the California Oroville–Thermalito complex. *Environ. Model. Softw.* **2015**, *69*, 262–279. [CrossRef]
14. Goyal, M.K.; Ojha, C.S.P.; Singh, R.D.; Swamee, P.K.; Nema, R.K. Application of ANN, fuzzy logic and decision tree algorithms for the development of reservoir operating rules. *Water Resour. Manag.* **2013**, *27*, 911–925.
15. Kim, T.; Heo, J.H.; Jeong, C.S. Multi-reservoir system optimization in the Han River basin using multi-objective genetic algorithms. *Hydrol. Process. Int. J.* **2006**, *20*, 2057–2075. [CrossRef]
16. Ming, B.; Chang, J.-X.; Huang, Q.; Wang, Y.-M.; Huang, S.-Z. Optimal Operation of Multi-Reservoir System Based-On Cuckoo Search Algorithm. *Water Resour. Manag.* **2015**, *29*, 5671–5687. [CrossRef]
17. Ostadrahimi, L.; Mariño, M.A.; Afshar, A. Multi-reservoir operation rules: Multi-swarm PSO-based optimization approach. *Water Resour. Manag.* **2012**, *26*, 407–427. [CrossRef]
18. Ming, B.; Liu, P.; Cheng, L.; Zhou, Y.; Wang, X. Optimal daily generation scheduling of large hydro–photovoltaic hybrid power plants. *Energy Convers. Manag.* **2018**, *171*, 528–540. [CrossRef]
19. Reddy, J.M.; Kumar, N.D. Multi-objective differential evolution with application to reservoir system optimization. *J. Comput. Civ. Eng.* **2007**, *21*, 136–146. [CrossRef]
20. Reddy, M.J.; Kumar, D.N. Performance evaluation of elitist-mutated multi-objective particle swarm optimization for integrated water resources management. *J. Hydroinform.* **2009**, *11*, 79–88. [CrossRef]
21. Wang, X.; Chang, J.; Meng, X.; Wang, Y. Research on multi-objective operation based on improved NSGA-II for lower Yellow River. *J. Hydraul. Eng.* **2017**, *48*, 135–145.

22. Yang, W. *Study on Multi-Objective Conversion Law of Joint Operation of Heihe Cascade Reservoir Group*; Xi'an University of Technology: Xi'an, China, 2018.
23. Nelsen, R.B. *An Introduction to Copulas*; Springer: New York, NY, USA, 2006.
24. Li, J.; Zhang, Q.; Chen, Y.D.; Singh, V.P. Future joint probability behaviors of precipitation extremes across China: Spatiotemporal patterns and implications for flood and drought hazards. *Planet. Chang.* **2015**, *124*, 107–122. [CrossRef]
25. Chen, Y.; Zhang, Q.; Xiao, M.; Singh, V.P.; Zhang, S. Probabilistic forecasting of seasonal droughts in the Pearl River basin, China. *Stoch. Environ. Res. Risk Assess.* **2016**, *30*, 2031–2040. [CrossRef]
26. Goswami, U.P.; Bhargav, K.; Hazra, B.; Goyal, M.K. Spatiotemporal and joint probability behavior of temperature extremes over the Himalayan region under changing climate. *Theor. Appl. Climatol.* **2018**, *134*, 477–498. [CrossRef]
27. Sraj, M.; Bezak, N.; Brilly, M. Bivariate flood frequency analysis using the copula function: A case study of the Litija station on the Sava River. *Hydrol. Process.* **2015**, *29*, 225–238. [CrossRef]
28. Huang, S.; Chang, J.; Huang, Q.; Chen, Y. Spatio-temporal changes and frequency analysis of drought in the Wei River Basin, China. *Water Resour. Manag.* **2014**, *28*, 3095–3110. [CrossRef]
29. Fang, W.; Huang, S.; Huang, G.; Huang, Q.; Wang, H.; Wang, L.; Zhang, Y.; Li, P.; Ma, L. Copulas-based risk analysis for inter-seasonal combinations of wet and dry conditions under a changing climate. *Int. J. Climatol.* **2019**, *39*, 2005–2021. [CrossRef]
30. Chen, F.; Huang, G.; Fan, Y.; Chen, J. A copula-based fuzzy chance-constrained programming model and its application to electric power generation systems planning. *Appl. Energy* **2017**, *187*, 291–309. [CrossRef]
31. Yu, L.; Li, Y.; Huang, G.; Fan, Y.; Nie, S. A copula-based flexible-stochastic programming method for planning regional energy system under multiple uncertainties: A case study of the urban agglomeration of Beijing and Tianjin. *Appl. Energy* **2018**, *210*, 60–74. [CrossRef]
32. Kong, X.M.; Huang, G.H.; Li, Y.P.; Fan, Y.R.; Zeng, X.T.; Zhu, Y. Inexact Copula-Based Stochastic Programming Method for Water Resources Management under Multiple Uncertainties. *J. Water Resour. Plan. Manag.* **2018**, *144*, 04018069. [CrossRef]
33. Fan, Y.R.; Huang, G.H.; Baetz, B.W.; Li, Y.P. Development of a copula-based particle filter (CopPF) approach for hydrologic data assimilation under consideration of parameter interdependence. *Water Resour.* **2017**, *53*, 4850–4875. [CrossRef]
34. Guo, S.; Yan, B.; Xiao, Y.; Fang, B.; Zhang, N. Application and research progress of Copula function in multivariate hydrological analysis and computation. *J. China Hydrol.* **2008**, *28*, 1–7.
35. Zhao, J.; Xu, Z.-X.; Zuo, D.-P.; Wang, X.-M. Temporal variations of reference evapotranspiration and its sensitivity to meteorological factors in Heihe River Basin, China. *Water Sci. Eng.* **2015**, *8*, 1–8. [CrossRef]
36. Min, L.; Yu, J.; Liu, C.; Zhu, J.; Wang, P. The spatial variability of streambed vertical hydraulic conductivity in an intermittent river, northwestern China. *Environ. Earth Sci.* **2013**, *69*, 873–883. [CrossRef]
37. Wang, Y.; Feng, Q.; Chen, L.; Yu, T. Significance and Effect of Ecological Rehabilitation Project in Inland River Basins in Northwest China. *Environ. Manag.* **2013**, *52*, 209–220.
38. Goldberg, D.E.; Deb, K. A Comparative Analysis of Selection Schemes Used in Genetic Algorithms. *Found. Genet. Algorithms* **1991**, *1*, 69–93.
39. Deb, K.; Agrawal, S.; Pratap, A.; Meyarivan, T. A fast elitist non-dominated sorting genetic algorithm for multi-objective optimization: NSGA-II. In Proceedings of the International Conference on Parallel Problem Solving from Nature, Paris, France, 18–20 September 2000; Springer: Berlin/Heidelberg, Germany, 2000; pp. 849–858.
40. Sklar, M. Fonctions de repartition an dimensions et leurs marges. *Publ. Inst. Statist. Univ. Paris* **1959**, *8*, 229–231.
41. Shiau, J.-T. Fitting Drought Duration and Severity with Two-Dimensional Copulas. *Water Resour. Manag.* **2006**, *20*, 795–815. [CrossRef]
42. Guo, Y.; Huang, S.; Huang, Q.; Wang, H.; Fang, W.; Yang, Y.; Wang, L. Assessing socioeconomic drought based on an improved Multivariate Standardized Reliability and Resilience Index. *J. Hydrol.* **2019**, *568*, 904–918. [CrossRef]
43. Yue, S.; Ouarda, T.; Bobée, B.; Legendre, P.; Bruneau, P. The Gumbel mixed model for flood frequency analysis. *J. Hydrol.* **1999**, *226*, 88–100. [CrossRef]

44. Zhao, M.; Huang, S.; Huang, Q.; Wang, H.; Leng, G.; Xie, Y. Assessing socio-economic drought evolution characteristics and their possible meteorological driving force. *Geomat. Hazards Risk* **2019**, *10*, 1084–1101. [CrossRef]
45. Guindon, S.; Gascuel, O. A Simple, Fast, and Accurate Algorithm to Estimate Large Phylogenies by Maximum Likelihood. *Syst. Biol.* **2003**, *52*, 696–704. [CrossRef]
46. Akaike, H. A new look at the statistical model identification. *IEEE Trans. Autom.* **1974**, *19*, 716–723. [CrossRef]
47. Yang, T.; Tao, Y.; Li, J.; Zhu, Q.; Su, L.; He, X.; Zhang, X. Multi-criterion model ensemble of CMIP5 surface air temperature over China. *Theor. Appl. Clim.* **2017**, *132*, 1057–1072. [CrossRef]

Case Report

Downscaling of SMAP Soil Moisture in the Lower Mekong River Basin

Chelsea Dandridge *, Bin Fang and Venkat Lakshmi

Department of Engineering Systems and Environment, University of Virginia, Charlottesville, VA 22904, USA; bf3fh@virginia.edu (B.F.); vlakshmi@virginia.edu (V.L.)

* Correspondence: cld9mt@virginia.edu; Tel.: +1-434-547-2207

Received: 8 November 2019; Accepted: 18 December 2019; Published: 21 December 2019

Abstract: In large river basins where in situ data were limited or absent, satellite-based soil moisture estimates can be used to supplement ground measurements for land and water resource management solutions. Consistent soil moisture estimation can aid in monitoring droughts, forecasting floods, monitoring crop productivity, and assisting weather forecasting. Satellite-based soil moisture estimates are readily available at the global scale but are provided at spatial scales that are relatively coarse for many hydrological modeling and decision-making purposes. Soil moisture data are obtained from NASA's soil moisture active passive (SMAP) mission radiometer as an interpolated product at 9 km gridded resolution. This study implements a soil moisture downscaling algorithm that was developed based on the relationship between daily temperature change and average soil moisture under varying vegetation conditions. It applies a look-up table using global land data assimilation system (GLDAS) soil moisture and surface temperature data, and advanced very high resolution radiometer (AVHRR) and moderate resolution imaging spectroradiometer (MODIS) normalized difference vegetation index (NDVI) and land surface temperature (LST). MODIS LST and NDVI are used to obtain downscaled soil moisture estimates. These estimates are then used to enhance the spatial resolution of soil moisture estimates from SMAP 9 km to 1 km. Soil moisture estimates at 1 km resolution are able to provide detailed information on the spatial distribution and pattern over the regions being analyzed. Higher resolution soil moisture data are needed for practical applications and modelling in large watersheds with limited in situ data, like in the Lower Mekong River Basin (LMB) in Southeast Asia. The 1 km soil moisture estimates can be applied directly to improve flood prediction and assessment as well as drought monitoring and agricultural productivity predictions for large river basins.

Keywords: SMAP; passive microwave soil moisture; soil moisture downscaling

1. Introduction

Estimating the water balance in large watersheds is of great interest for water resource management and soil moisture is a key variable in this estimation as it effects evaporation, infiltration, and runoff [1]. Soil moisture acts as a link between energy and water fluxes at Earth's surface-atmosphere interface, and knowledge of soil moisture variation is the key to understanding the hydrological cycle [2]. Soil moisture is the primary source of water for agriculture and directly influences crop growth and food production [3]. Even though it only accounts for a small portion of global freshwater, it is still an important factor in global hydrologic cycles [3]. This seemingly small layer (top few centimeters) controls the regulation and distribution of precipitation between runoff and water storage [4]. Soil moisture observations over large areas are increasingly necessary for a range of applications such as meteorology, hydrology, water resource management, and climatology [5]. Remote sensing has provided valuable data sets for understanding land surface hydrological and meteorological processes [6–9].

Obtaining soil moisture measurements can be achieved using a variety of remote sensing instruments or ground-based systems. Satellite-based radars can measure soil moisture at high resolution but are limited in spatial coverage and temporal frequency. Satellite data products can produce global soil moisture estimates but are usually too coarse for practical use in modelling and decision-making [10]. High resolution soil moisture estimates can be applied directly to improve flood prediction and assessment as well as drought monitoring, agricultural productivity prediction, and irrigation management [11–14]. With improved prediction of extreme events, we can also better prepare for their effects on the natural environment and future climate change [2]. NASA's soil moisture active passive (SMAP) will help determine whether there will be more or less water, regionally, in the future compared to today [15,16]. Monitoring these changes in future water resources is a very important aspect of climate change as this will affect the future water supply and food production in areas like the Lower Mekong Basin [17–19]. High resolution soil moisture can aid in crop yield forecasting as well as by providing earlier monitoring of droughts and better understanding of hydrologic processes [4].

This research uses global soil moisture data derived from the L-band radiometer aboard NASA's SMAP observatory [20]. However, satellite microwave radiometers are much coarser than active microwave and optical systems [6]. This coarseness reduces satellite applicability in large watershed models and for regional flood prediction [21]. This study aims to downscale SMAP soil moisture estimates, from gridded 9 km resolution to 1 km resolution, in the Lower Mekong Basin (LMB). This will be done using the regression relationship between daily temperature changes and daily soil moisture under different vegetation conditions with the algorithm developed by Fang et al., 2013. Soil moisture estimates with high spatial resolution can be very useful for watershed scale hydrological modeling due to the fact that soil moisture estimates can be used to constrain errors during extensive wetting and dry downs [21]. The downscaling algorithm and methodology implemented in this research were developed in a previous study by Fang et al., 2018. This algorithm has been applied to the Black Bear-Red Rock watershed in Oklahoma and validated with in situ soil moisture from the ISMN (International Soil Moisture Network). Regions with low elevation are vulnerable to flooding and other water-resource related problems. With these problems, it is important to increase the capacity of flood and drought monitoring. Here we apply this validated algorithm to the Lower Mekong Basin, an area with no functioning in situ soil moisture network. With higher resolution soil moisture, this region would have greater modelling capabilities and the ability to make better decisions concerning water resource management. This algorithm can be applied to other watersheds worldwide, with little absent from the in situ soil moisture systems.

The Mekong River in Southeast Asia provides food, water, and energy resources to the countries of China, Laos, Myanmar, Thailand, Cambodia, and Vietnam [2]. It is the 12th longest river in the world, extending over 4300 km [22]. The basin can be divided into two major catchments also known as the upper and lower river basins. The upper basin is mostly mountainous, rising in the Tibetan Plateau (Figure 1). The Lower Mekong Basin (LMB) is subject to high levels of flooding due to the combination of low-lying terrain and seasonal precipitation cycles [22]. The LMB is home to the rice paddy fields of Vietnam, which would benefit greatly from consistent soil moisture data. Unfortunately, the LMB does not have a consistent in situ soil moisture measuring system, which makes satellite-derived soil moisture estimates appealing for application in watershed-scale hydrological modelling in this region. The lack of ground measurements for soil moisture also complicated the validity of remotely-sensed estimates of the LMB [2].

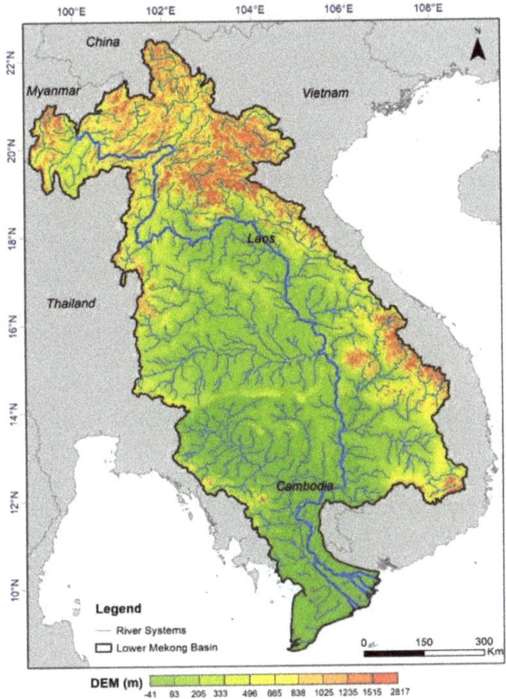

Figure 1. Topography and river networks in Lower Mekong River Basin (LMB).

2. Data

2.1. SMAP Data

Developed by NASA, the soil moisture active passive (SMAP) observatory was designed to distinguish between frozen and thawed land surfaces [14]. This mission was launched in January 2015 with the goal of combining radar and radiometer at L-band frequencies to record high resolution soil moisture measurements and freeze/thaw detection at global scale. Unfortunately, shortly after the launch a hardware failure caused the radar to stop working, leaving the radiometer as the only operational mechanism to record data [23]. Since the launch, the radiometer aboard the observatory has been collecting data at a spatial resolution of 36 km and providing global coverage every 2 to 3 days [23]. Observations from SMAP will provide improved estimates of water, energy, and transfers between land and atmosphere [24,25]. SMAP uses lower frequency microwave radiometry (L Band) to map soil moisture at Earth's land surface because at lower frequencies the atmosphere is less opaque, vegetation is more transparent, and the results were more representative of the soil below the skin surface than when higher frequencies were used [26,27]. This research utilizes the SMAP Level 2 enhanced passive soil moisture product (L2_SM_P_E), which is available on a 9-km grid for downscaling to 1-km resolution.

2.2. GLDAS Data

NASA's global land data assimilation system (GLDAS) was designed to combine satellite- and observation-based data to produce high resolution, global information on Earth's land surface states and fluxes [28]. GLDAS is able to provide 36 land surface fields from 2000 to the present, including soil moisture, surface temperature, surface runoff, and rainfall. The product of 3-hourly data (GLDAS_NOAH025_3H) with $0.25° \times 0.25°$ spatial resolution was used in this study [29].

Our downscaling approach utilized soil moisture with 0 to 10 cm depth and surface skin temperature from GLDAS that corresponded to the closest overpass times of the Aqua satellite for the LMB, which was approximately 12:00 and 24:00 local time.

2.3. MODIS Data

NASA's moderate resolution imaging spectroradiometer (MODIS) was launched aboard the Earth observing system (EOS) aqua satellite in May 2002 and provides atmospheric, terrestrial and oceanic data products [30]. With 36 spectral bands, the highest of any global coverage moderate resolution imager, and spatial resolution ranging from 250 m to 1 km, MODIS is able to provide a multitude of global land products [30]. In this study, daily normalized difference vegetation index (NDVI) and land surface temperature (LST) from MODIS were used to downscale SMAP soil moisture estimates. The 1 km daily LST (MYD11A1), 1 km biweekly NDVI (MYD13A2), and 500 m biweekly climate modeling grid (CMG) NDVI (MYD13C1) were utilized in this study.

2.4. AVHRR Data

The advanced very high resolution radiometer (AVHRR) utilizes National Oceanic and Atmospheric Administration (NOAA) polar-orbiting satellites to provide four- to six- band multispectral global data [31]. The AVHRR is used to remotely detect cloud cover and the Earth's surface temperature (NOAA satellite information system, 2013). Prior to MODIS data, AVHRR's 5 km CMG NDVI data were used for long-term surface ground measurements [11]. In this study, daily NDVI data (AVH13C1) from AVHRR from 1981 to 1999 were used. The quality of AVHRR data after this time period is inadequate due to satellite drifting and, therefore, data after 2000 was not used in this study [11] (Table 1).

Table 1. Description of the data products used in the downscaling process including their spatial and temporal resolutions and data availability.

Data Product	Variable	Spatial Resolution	Temporal Resolution	Availability
SMAP	Soil moisture	9 km	Daily	2015–present
GPM IMERG	Precipitation	10 km	Daily	2000–present
GLDAS	Soil moisture	25 km	3 hours	1979–present
MODIS	Land surface temperature (LST)	1 km	Daily	2002–present
MODIS	NDVI	1 km	Biweekly	2002–present
AVHRR	NDVI	5 km	Daily	1981–1999

3. Methodology

In this research, the downscaling algorithm and methodology used were developed in Fang et al., 2018 [32]. Similar to the study by Lakshmi and Fang (2015) of the Little Washita Watershed, this study assumes that LST is a linear combination of soil and vegetation temperature [33]. We assume the top soil moisture layer is a function of soil evaporation efficiency and field capacity. It is assumed that the soil moisture at a certain time during the day is inversely proportional to the daily temperature change for the same day, and that the presence of vegetation (NDVI) will influence the soil moisture–temperature change relationship. We also assume that the thermal inertia relationship between temperature difference and soil moisture within a 25 km domain has no spatial variability. Additionally, the assumption is made that the field capacity of each NLDAS pixel is homogenous and does not account for variation at the 1 km scale [32].

In this study, we applied an algorithm developed by Fang et al., 2018, based on soil moisture, LST, and NDVI, to create 1 km soil moisture maps [32]. The methodology of this algorithm is outlined in Figure 2. Due to the effects of vegetation cover on soil moisture estimation, the algorithm applied here uses a vegetation-based lookup table to relate microwave polarization to soil moisture estimates. As soil becomes more wet its heat capacity increases. The soil moisture at a given time is inversely proportional to the change in temperature 12 hours beforehand, which corresponds with SMAP AM and PM overpasses. Soil moisture daily values were negatively related to the daily temperature difference under varying vegetation conditions. The following equation represents the linear relationship between soil moisture and temperature difference for a specific NDVI (single month):

$$\theta(i,j) = a_0 + a_1 \Delta T_s(i,j) \tag{1}$$

where $\theta(i,j)$ is GLDAS soil moisture gridded to match SMAP overpasses and $\Delta T_s(i,j)$ is the GLDAS 12 h temperature difference closest and prior to SMAP overpasses. This equation uses data at the GLDAS spatial resolution for soil moisture and surface temperature for single months, beginning in 1981. Using the nearest neighbor method, daily NDVI from AVHRR was aggregated to corresponding GLDAS pixels. The NDVI data were categorized into classes from 0 to 1 with increments at 0.1. Classes with less than 8 data points were not included because a sample size smaller than this will not yield valid and statistically significant results from linear regression fitting. Soil moisture at 1 km resolution was calculated from 1 km MODIS LST difference at the corresponding NDVI class. We applied the linear regression fit equation between θ and ΔT_s, which was built at 25 km resolution, to all the 1 km MODIS grids within the 25 km GLDAS grid. We assumed that the thermal inertia relationship between temperature difference and soil moisture within the 25 km domain had no spatial variability. The following equation represents the correction of the 1 km soil moisture pixel from the MODIS LST products, acquired by removing the difference between SMAP and MODIS derived soil moisture:

$$\theta^{corr}(i,j) = \theta(i,j) + \left[\Theta - \frac{1}{n}\sum_{i=1}^{n}\Theta_i\right] \tag{2}$$

where $\theta^{corr}(i,j)$ is the corrected 1 km soil moisture, n is the number of 1 km soil moisture pixels that are in each SMAP 9 km pixel, Θ is the original SMAP 9 km soil moisture estimate, and θ_i is the number of uncorrected 1 km SMAP soil moisture pixels that fall in the original 9 km SMAP grid Θ. The value of n is ideally 81, but it may be less due to cloud contaminated data. The corrected soil moisture was characterized by the soil moisture and daily temperature relationship, which changed under different vegetation conditions. Since visualizing rainfall is essential to determining the response of the soil moisture, rainfall from GPM IMERG was used in this study to analyze the wetting and dry-down patterns after a significant rainfall event. One limitation of this methodology occurred when the 9 km original SMAP was biased. Then that bias was passed onto the corrected 1 km soil moisture. Another limitation was the difficulty to recover cloud-contaminated data, which resulted in spatial inconsistencies in the 1 km corrected soil moisture maps.

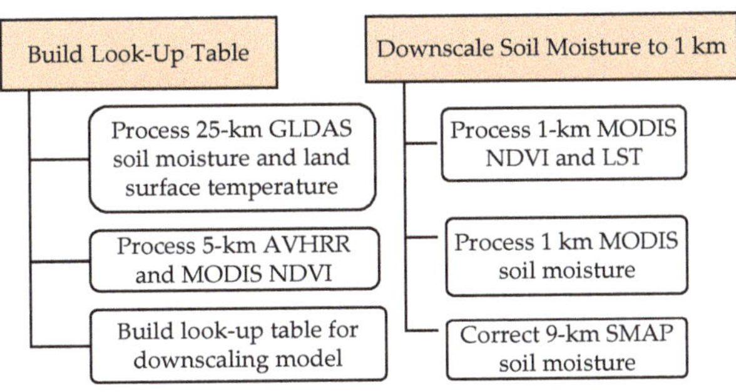

Figure 2. Workflow for building downscaling model and executing the algorithm.

The algorithm used in this study was validated using in situ measurements in the CONUS region, by Fang et al., 2018, for soil moisture estimates from AMSR2 between 2015 and 2017. Their validation showed variability in seasonal performance and stronger correlations in the soil moisture–temperature change relationship during summer months. Also, the remotely sensed soil moisture and downscaled estimates both underestimated in situ soil moisture during precipitation events. It is important to note the effects of precipitation on soil moisture retrieval; the microwave sensing depth is reduced. An additional validation of this algorithm was performed in the Walnut Gulch Experimental Watershed (WGEW) and indicated that downscaled soil moisture had better validation metrics than the original SMAP [32]. The R^2 of the 1 km soil moisture ranged from 0.189 to 0.697, whereas the 9 km SMAP ranged from 0.003 to 0.597. The slope values for the 1 km are higher than those for the 9 km SMAP. Additionally, the 1 km soil moisture RMSE values and biases improved compared to the original SMAP data. There were no consistent soil moisture measurements in the Lower Mekong Basin, and this presents a formidable challenge to validation. However, future work may be able to carry out validation by comparison of the 1 km soil moisture to outputs from hydrological models.

4. Results

4.1. Rainfall Variation in the Lower Mekong Basin

Variations in rainfall patterns result in changes in soil moisture. Precipitation has a direct impact on the wetting and drying of soils and, therefore, must be examined alongside soil moisture. In the LMB, the annual wet season (April–September) results in more vegetation growth and cloud cover compared to the dry season. Therefore, the ability to measure soil moisture via remote sensing is affected during these months. Daily precipitation data from GPM IMERG Final Precipitation L3 1 day 0.1° by 0.1° V05 (GPM_3IMERGDF) were aggregated for monthly accumulation for April through September from 2015 to 2018, to correspond with the downscaled soil moisture in order to examine the monthly variations (Figure 3) [34].

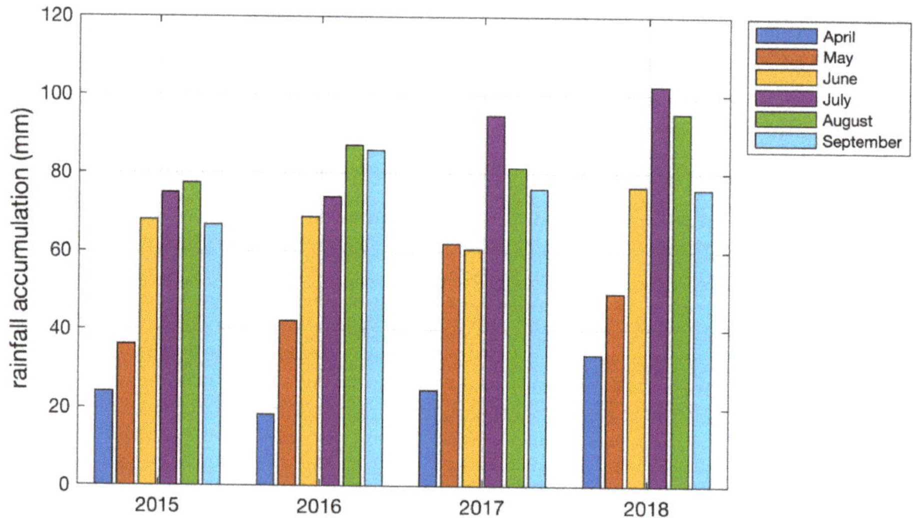

Figure 3. Bar plot of monthly average precipitation for April 2015–September 2018 in the LMB.

In this study, precipitation patterns varied in the wet season months, with July and August generally accumulating the most rainfall and April and May receiving the least (Figure 3). Additionally, precipitation varied from year to year over the LMB, with certain years being more dry or wet than others due to regulation by monsoons. For example, comparing the year 2016 to 2018 in Figure 3 shows 2016 as a much dryer year, especially in the wettest month of the year, July, which received over 100 mm of rainfall. This pattern can also be seen by comparing the monthly maps from 2016 and 2018 (Figure 4). Figure 4 shows the spatial distribution of accumulated precipitation over the LMB for each month, corresponding to the 1 km soil moisture estimates. Rainfall patterns varied significantly between countries in the LMB. Areas in Laos and Cambodia receive the greatest amounts of precipitation annually (over 2800 mm), while the Thailand plateau only received a third or less of that amount. Here, precipitation from IMERG was used to detect and observe the dry-down patterns of soil moisture after a large rainfall event.

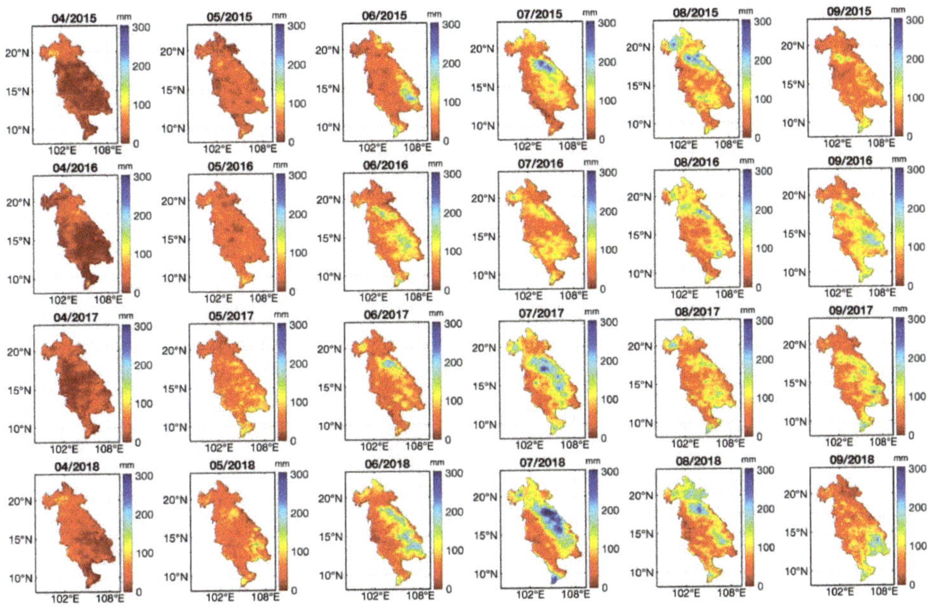

Figure 4. Monthly rainfall accumulation from GPM IMERG for April 2015–September 2018 in the LMB.

4.2. Soil Dryness Response to Large Rainfall Events

In this section, soil moisture is examined alongside precipitation with the purpose of examining the drying of soil over time in response to a rainfall event. By evaluating the time series after a large precipitation event with almost no subsequent precipitation, we were able to observe the near-surface soil moisture observations as they transitioned from saturated to dry conditions. Daily 9 km SMAP soil moisture estimates were compared to daily 10 km IMERG precipitation to examine the response of soil moisture to precipitation events. It is possible that, in the absence of precipitation, agriculture is irrigated. Hence, we may have seen wetness from irrigation in these regions, despite no significant rainfall event. Figure 5 shows the relationship between daily rainfall and soil moisture between 2015 and 2018 averaged over the LMB.

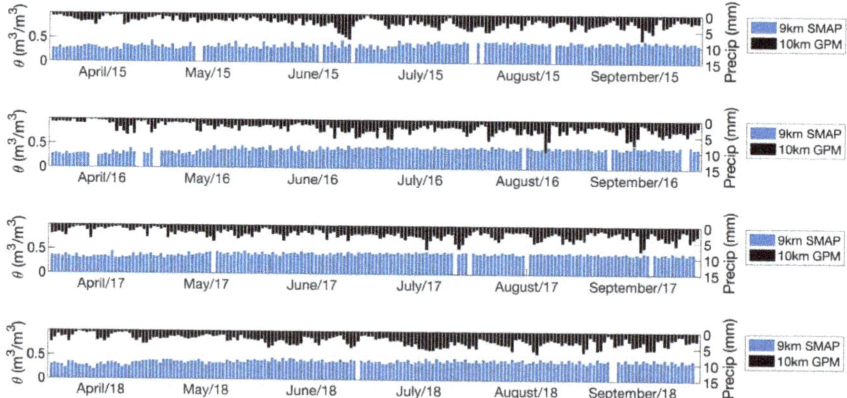

Figure 5. Time series of daily soil moisture active passive (SMAP) 9 km soil moisture and daily Global Precipitation Measurement-Integrated Multi-satellitE Retrieval (GPM-IMERG) 10 km precipitation for April 2015–September 2018 averaged in the LMB.

Using Figure 5, two precipitation events were selected in which soil moisture exhibited a clear dry-down pattern after the rainfall. The events were examined more closely in combination with corresponding daily downscaled soil moisture, in order to evaluate the improvement in the representation of drying from 9 km to 1 km. Figure 6 more closely examines the time series of the dry-down period in the LMB from 13 April 2015 to 20 April 2015, after a large precipitation event occurred on 13 April. Figure 7 shows the spatial distribution of rainfall, 9 km SMAP soil moisture, and 1 km downscaled soil moisture for each day during the dry-down period. The second event selected was from 6 April 2018 to 11 April 2018. Figure 8 shows the time series of the dry-down period after the precipitation event on 6 April 2018. The 1 km soil moisture (blue) was better able to capture the dry-down pattern than the 9 km SMAP soil moisture (green) (Figure 8). Figure 9 shows the spatial distribution of rainfall, 9 km SMAP soil moisture, and 1 km downscaled soil moisture for each day during the dry-down period in April 2018. The coverage of the 1 km corrected soil moisture was dependent on MODIS LST data and influenced by cloud cover, which made it difficult to find good coverage on consecutive days. The 1 km SMAP did not perform as well during wet days due to the spatial coverage of the MODIS land surface temperature (LST) data being compromised by cloud contamination.

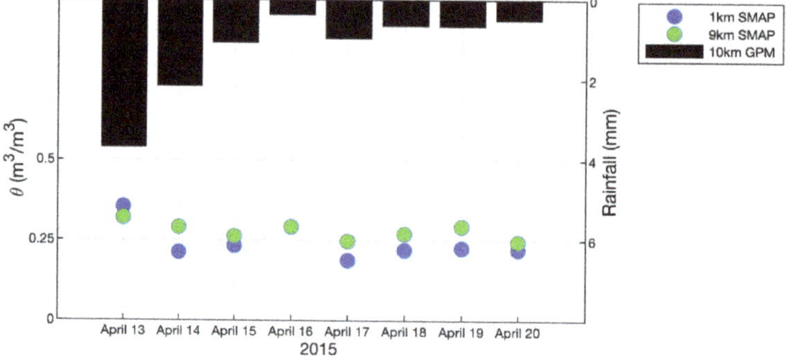

Figure 6. Time series of dry-down event from 13 April 2015 to 20 April 2015 with 9 km SMAP soil moisture (green), 1 km downscaled soil moisture (blue), and 10 km rainfall data from GPM IMERG (black).

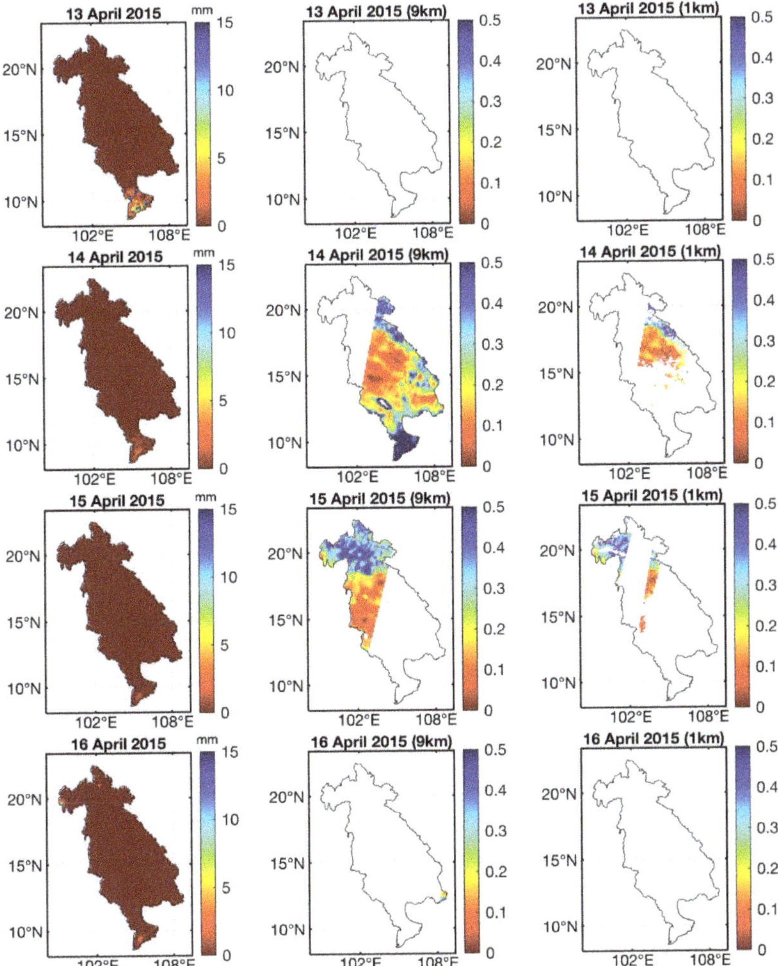

Figure 7. *Cont.*

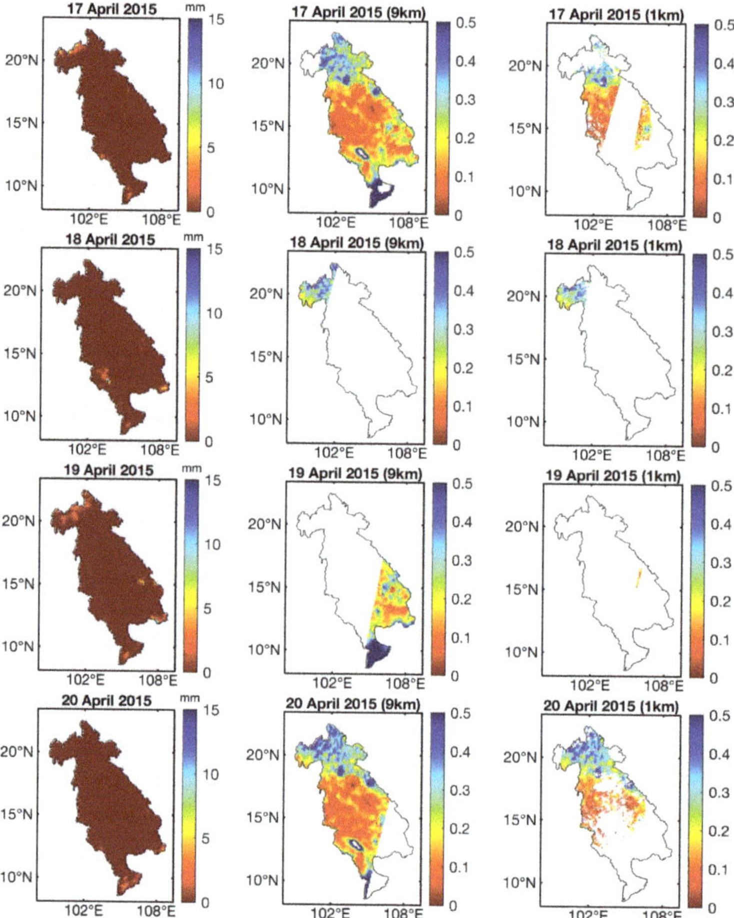

Figure 7. Dry-down event for 13 April 2015 to 20 April 2018 represented by 10 km IMERG rainfall, 9 km SMAP, and 1 km downscaled soil moisture.

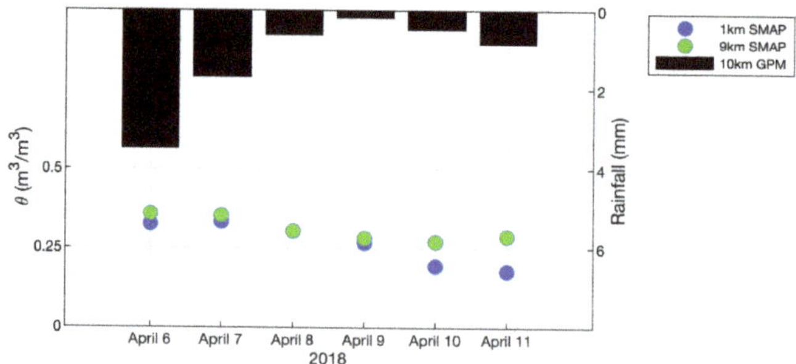

Figure 8. Time series of dry-down event from 6 April 2018 to 11 April 2018 with 9 km SMAP soil moisture (green), 1 km downscaled soil moisture (blue), and 10 km rainfall data from GPM IMERG (black).

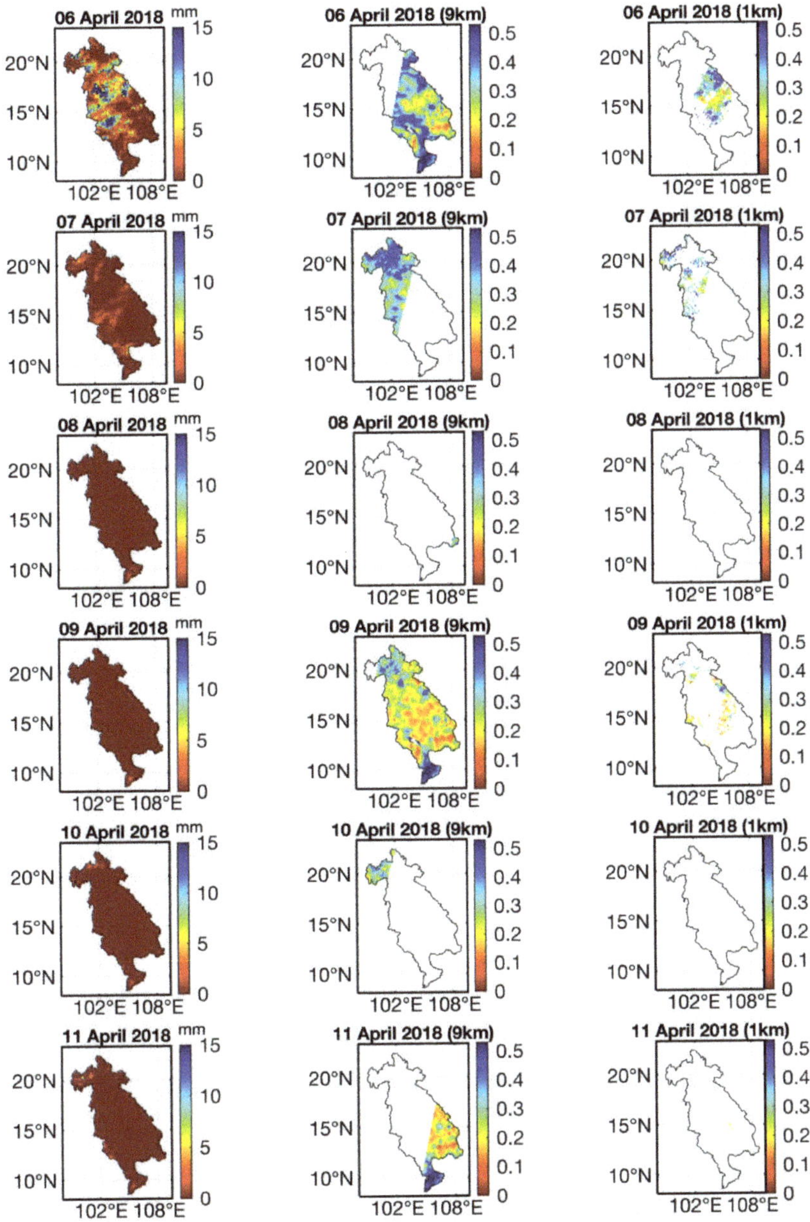

Figure 9. Dry-down event for 6 April 2018 to 11 April 2018 represented by 10 km IMERG rainfall, 9 km SMAP, and 1 km downscaled soil moisture.

4.3. Importance of High Spatial Resolution Soil Moisture for Hydrology and Water Resources

The high spatial resolution observed soil moisture generated in this study was an important data set that could not be obtained from other sources. Firstly, there are no consistent in situ networks that monitor soil moisture in the Lower Mekong River Basin. Even in other parts of the world that do have such networks, they are seldom dense enough to produce soil moisture at 1 km spatial resolution. Secondly, although land surface models can simulate soil moisture at high spatial resolution, they lack the precipitation input at 1 km spatial resolution, which is needed to minimize variations in small-scale processes [35]. Currently, the "best" spatial resolution of globally available precipitation is the climate hazards group infrared precipitation with station observations (CHIRPS) at 0.05°. CHIRPS provides estimates from 1981 to the near present and uses a recently produced satellite rainfall algorithm that combines climatology data, satellite precipitation estimates, and in situ rain gauge measurements to produce a high resolution precipitation product [36].

The 1 km spatial resolution soil moisture from this research can be used in combination with land use and land cover data from MODIS (moderate resolution imaging spectroradiometer) at 1 km and Landsat imagery at 30 m to map the co-variability of land use and wetness. This will be a valuable tool for land use planning, specifically in the LMB where there are competing cropping strategies and land use for industrial development. Additionally, this 1 km soil moisture can be used to determine antecedent soil moisture conditions in watershed modeling, meaning it can serve as an input to determine the portion of rainfall that will infiltrate the soil and that which will run off to the stream network. More detailed estimations of streamflow runoff will in turn benefit flood prediction and monitoring in watersheds [37]. This high spatial resolution 1 km observed soil moisture can serve a variety of water resource applications and will be of much use in the LMB.

5. Conclusions

This study applied a previously developed method to a new geographical location where in situ observations are lacking. Here, higher resolution could help various land use decisions such as construction of dams, agriculture, and aquaculture. In this study, soil moisture estimates of the Lower Mekong River Basin from April 2015–September 2018, from SMAP Enhanced L2 Radiometer Half-Orbit 9 km V.2., were enhanced to 1 km resolution. In this study, we applied an algorithm developed by Fang et al., 2018, based on soil moisture, LST, and NDVI to create 1 km soil moisture maps. Soil moisture daily values were negatively related to the daily temperature difference under varying vegetation conditions. The downscaling algorithm was based on LST, soil moisture, and NDVI and used the relationship between daily soil moisture and daily land surface temperature difference between satellite overpasses as well as the vegetation class to downscale soil moisture to a higher resolution. The months of April and May showed the best coverage of soil moisture at 1 km and July–September showed the least coverage at 1 km, due to LST/NDVI data with substantial cloud coverage and higher vegetation growth. It was discovered in this study that the 1 km SMAP did not perform as well during wet days due to the spatial coverage of the MODIS land surface temperature (LST) data being compromised by cloud contamination.

Soil moisture estimates are readily available at global scale from a multitude of satellite products but are represented at spatial scales that are often too coarse for effective hydrological modeling and decision-making purposes. Soil moisture at high resolution can be used in place of ground measurements for land and water management decisions in large river basins where in situ data are limited such as the LMB. The high resolution soil moisture estimates derived in this study can be more useful for assessing dry-down and wetting trends than coarser resolution data, such as the 9 km SMAP product in the LMB. Additionally, 1 km soil moisture retrievals can better aid drought and crop productivity monitoring, flood forecasting, and assist weather forecasting by providing greater spatial representation than coarser products. This high spatial resolution soil moisture at 1 km can be applied to a multitude of water resources applications in order to benefit large watershed management.

Author Contributions: Individual contributions from the authors include conceptualization, C.D.; B.F.; V.L.; methodology, B.F.; software, C.D.; validation, B.F.; V.L.; formal analysis, C.D.; B.F.; investigation, C.D.; resources, V.L.; data curation, C.D.; and B.F.; writing—original draft preparation, C.D.; writing—review and editing, C.D.; B.F.; V.L.; visualization C.D.; B.F.; V.L.; supervision, V.L.; project administration, V.L. All authors have read and agreed to the published version of the manuscript.

Funding: This research received no external funding.

Conflicts of Interest: The authors declare no conflict of interest.

References

1. Renzullo, L.; van Dijk, A.; Perraud, J.-M.; Collins, D.; Henderson, B.; Jin, H.; Smith, A.; McJannet, D. Continental satellite soil moisture data assimilation improves root-zone moisture analysis for water resources assessment. *J. Hydrol.* **2014**. [CrossRef]
2. Naeimi, V.; Leinenkugel, P.; Sabel, D.; Wagner, W.; Apel, H.; Kuenzer, C. Evaluation of soil moisture retrieval from the ERS and metop scatterometers in the lower mekong basin. *Remote Sens.* **2013**, *5*, 1603–1623. [CrossRef]
3. Dobriyal, P.; Qureshi, A.; Badola, R.; Hussain, S.A. A review of the methods available for estimating soil moisture and its implications for water resource management. *J. Hydrol.* **2012**, *458*, 110–117. [CrossRef]
4. Engman, E.T. Application of microwave remote sensing of soil moisture for water resources and agriculture. *Remote Sens. Environ.* **1991**, *35*, 213–226. [CrossRef]
5. O'Neill, P.E.; Chan, S.; Njoku, E.G.; Jackson, T.; Bindlish, R.; Chaubell, J. *SMAP Enhanced L2 Radiometer Half-Orbit 9 km EASE-Grid Soil Moisture, Version 3*; NASA National Snow and Ice Data Center, Distributed Active Archive Center: Boulder, CO, USA, 2017; Available online: https://nsidc.org/data/SPL2SMP_E/versions/3 (accessed on 13 August 2019). [CrossRef]
6. Merlin, O.; Walker, J.; Chehbouni, A.; Kerr, Y. Towards deterministic downscaling of SMOS soil moisture using MODIS derived soil evaporative efficiency. *Remote Sens. Environ.* **2008**, *112*, 3935–3946. [CrossRef]
7. Lakshmi, V.; Small, E.E.; Hong, S.; Chen, F. The influence of the land surface on hydrometeorology and ecology: New advances from modeling and satellite remote sensing. *Hydrol. Res.* **2011**, *42*, 95–112. [CrossRef]
8. Lakshmi, V. The role of remote sensing in prediction of ungaged basins. *Hydrol. Process.* **2004**, *18*, 1029–1034. [CrossRef]
9. Hong, S.; Lakshmi, V.; Small, E. Relationship between vegetation biophysical properties and surface temperature: Results using satellite data. *J. Clim.* **2007**, *20*, 5593–5606. [CrossRef]
10. Billah, M.M.; Goodall, J.; Narayan, U.; Reager, J.; Lakshmi, V.; Famiglietti, J. A methodology for evaluating evapotranspiration estimates at the watershed-scale using GRACE. *J. Hydrol.* **2015**, *523*, 574–586. [CrossRef]
11. Fang, B.; Lakshmi, V.; Bindlish, R.; Jackson, T.J.; Cosh, M.; Basara, J. Passive microwave soil moisture Downscaling using vegetation index and skin surface temperature. *Vadose Zone J.* **2013**, *12*, 3. [CrossRef]
12. Schmugge, T.J.; Kustas, W.P.; Ritchie, J.C.; Jackson, T.J.; Rango, A. Remote sensing in hydrology. *Adv. Water Resour.* **2002**, *25*, 1367–1385. [CrossRef]
13. Wagner, W.; Blöschl, G.; Pampaloni, P.; Calvet, J.C.; Bizzarri, B.; Wigneron, J.P.; Kerr, Y. Operational readiness of microwave remote sensing of soil moisture for hydrologic applications. *Hydrol. Res.* **2007**, *38*, 1–20. [CrossRef]
14. Entekhabi, D.; Njoku, E.; Oneill, P.; Spencer, M.; Jackson, T.; Entin, J.; Jhonson, J.; Kimbal, J.; Kellogg, K.; Martin, N.; et al. The soil moisture active/passive mission (SMAP). *Int. Geosci. Remote Sens. Symp.* **2010**, *98*, 704–716. [CrossRef]
15. Abbaszadeh, P.; Moradkhani, H.; Zhan, X. Downscaling SMAP radiometer soil moisture over the CONUS using an ensemble learning method. *Water Resour. Res.* **2019**, *55*, 324–344. [CrossRef]
16. McNairn, H.; Jackson, T.J.; Wiseman, G.; Belair, S.; Berg, A.; Bullock, P.; Shang, J.L.; Hosseini, M.; Rowlandson, T.L.; Moghaddam, M.; et al. The soil moisture active passive validation experiment 2012 (SMAPVEX12): Prelaunch calibration and validation of the SMAP soil moisture algorithms. *IEEE Trans. Geosci. Remote Sens.* **2014**, *53*, 2784–2801. [CrossRef]
17. Mohanty, B.P.; Cosh, M.H.; Lakshmi, V.; Montzka, C. Soil moisture remote sensing: State-of-the-science. *Vadose Zone J.* **2017**, *16*. [CrossRef]
18. Delworth, T.; Manabe, S. The influence of soil wetness on near-surface atmospheric variability. *J. Clim.* **1989**, *2*, 1447–1462. [CrossRef]
19. Brubaker, K.L.; Entekhabi, D. Analysis of feedback mechanisms in land-atmosphere interaction. *Water Resour. Res.* **1996**, *32*, 1343–1357. [CrossRef]
20. Lakshmi, V. Remote sensing of soil moisture. *ISRN Soil Sci.* **2013**, *2013*, 33. [CrossRef]

21. Narayan, U.; Lakshmi, V. Characterizing subpixel variability of low resolution radiometer derived soil moisture using high resolution radar data. *Water Resour. Res.* **2008**, *44*, 6. [CrossRef]
22. Oddo, P.; Ahamed, A.; Bolten, J. Socioeconomic impact evaluation for near real-time flood detection in the lower mekong river basin. *Hydrology* **2018**, *5*, 23. [CrossRef]
23. Chan, S.K.; Bindlish, R.; O'Neill, P.; Jackson, T.; Njoku, E.; Dunbar, S.; Kerr, Y.; Berg, A.; Chen, F.; Walker, J.; et al. Development and assessment of the SMAP enhanced passive soil moisture product. *Remote Sens. Environ.* **2018**, *204*, 931–941. [CrossRef]
24. Peng, J.; Loew, A.; Merlin, O.; Verhoest, N.E. A review of spatial downscaling of satellite remotely sensed soil moisture. *Rev. Geophys.* **2017**, *55*, 341–366. [CrossRef]
25. Wigneron, J.P.; Jackson, T.J.; O'neill, P.; De Lannoy, G.; De Rosnay, P.; Walker, J.P.; Kerr, Y.; Royer, A.; Das, N.; Kurum, M.; et al. Modelling the passive microwave signature from land surfaces: A review of recent results and application to the L-band SMOS & SMAP soil moisture retrieval algorithms. *Remote Sens. Environ.* **2017**, *192*, 238–262.
26. Panciera, R.; Walker, J.P.; Jackson, T.J.; Gray, D.A.; Tanase, M.A.; Ryu, D.; Wu, X.L.; Yardley, H.; Monerris, A.; Gao, Y.; et al. The soil moisture active passive experiments (SMAPEx): Toward soil moisture retrieval from the SMAP mission. *IEEE Trans. Geosci. Remote Sens.* **2013**, *52*, 490–507. [CrossRef]
27. Colliander, A.; Jackson, T.J.; Bindlish, R.; Chan, S.; Das, N.; Kim, S.B.; Asanuma, J.; Aida, K.; Berg, A.; Bosch, D.; et al. Validation of SMAP surface soil moisture products with core validation sites. *Remote Sens. Environ.* **2017**, *191*, 215–231. [CrossRef]
28. Rodell, M.; Houser, P.R.; Jambor, U.; Gottschalck, J.; Mitchell, K.; Meng, C.-J.; Arsenault, K.; Cosgrove, B.; Radakovich, J.; Bosilovich, M.; et al. The global land data assimilation system. *Bull. Am. Meteorol. Soc.* **2004**, *85*, 381–394. [CrossRef]
29. Beaudoing, H.; Rodell, M. NASA/GSFC/HSL GLDAS noah land surface model L4 3 hourly 0.25 × 0.25 degree V2.1, greenbelt, maryland, USA. *Goddard Earth Sci. Data Inf. Serv. Cent.* **2018**, *10*, 5067.
30. Justice, C.; Townshend, J.; Vermote, E.; Masuoka, E.; Wolfe, R.; Saleous, N.; Roy, D.; Morisette, J. An overview of MODIS land data processing and product status. *Remote Sens. Environ.* **2002**, *83*, 3–15. [CrossRef]
31. Advanced Very High Resolution Radiometer (AVHRR). Available online: https://lta.cr.usgs.gov/AVHRR (accessed on 31 July 2018).
32. Fang, B.; Lakshmi, V.; Bindlish, R.; Jackson, T. Amsr2 soil moisture downscaling using temperature and vegetation data. *Remote Sens.* **2018**, *10*, 1575. [CrossRef]
33. Lakshmi, V. *Remote Sensing of the Terrestrial Water Cycle*; John Wiley & Sons: Washington, WA, USA, 2015; pp. 277–304.
34. Huffman, G.J.; Stocker, E.F.; Bolvin, D.T.; Nelkin, E.J.; Tan, J. GPM IMERG Final Precipitation L3 1 Day 0.1 Degree × 0.1 Degree V06 (Dataset). Goddard Earth Sciences Data and Information Services Center (GES DISC). Available online: https://disc.gsfc.nasa.gov/datacollection/GPM_3IMERGDF_06 (accessed on 5 September 2019).
35. Ochoa-Rodriguez, S.; Wang, L.-P.; Gires, A.; Pina, R.D.; Reinoso-Rondinel, R.; Bruni, G.; Veldhuis, M.-C.T. Impact of spatial and temporal resolution of rainfall inputs on urban hydrodynamic modelling outputs: A multi-catchment investigation. *J. Hydrol.* **2015**, *531*, 389–407. [CrossRef]
36. Dinku, T.; Funk, C.; Peterson, P.; Maidment, R.; Tadesse, T.; Gadain, H.; Ceccato, P. Validation of the CHIRPS satellite rainfall estimates over eastern Africa. *J. R. Meteorol. Soc.* **2018**, *144*, 292–312. [CrossRef]
37. Alfieri, L.; Burek, P.; Dutra, E.; Krzeminski, B.; Muraro, D.; Thielen, J.; Pappenberger, F. GloFAS–global ensemble streamflow forecasting and flood early warning. *Hydrol. Earth Syst. Sci.* **2013**, *17*, 1161–1175. [CrossRef]

MDPI
St. Alban-Anlage 66
4052 Basel
Switzerland
Tel. +41 61 683 77 34
Fax +41 61 302 89 18
www.mdpi.com

Water Editorial Office
E-mail: water@mdpi.com
www.mdpi.com/journal/water

www.ingramcontent.com/pod-product-compliance
Lightning Source LLC
LaVergne TN
LVHW070926170726
843515LV00005B/880